SpringerBriefs on PDEs and Data Science

Editor-in-Chief

Enrique Zuazua, Department of Mathematics, University of Erlangen-Nuremberg, Erlangen, Bayern, Germany

Series Editors

Irene Fonseca, Department of Mathematical Sciences, Carnegie Mellon University, Pittsburgh, PA, USA

Franca Hoffmann, Hausdorff Center for Mathematics, University of Bonn, Bonn, Germany

Shi Jin, Institute of Natural Sciences, Shanghai Jiao Tong University, Shanghai, Shanghai, China

Juan J. Manfredi, Department of Mathematics, University Pittsburgh, Pittsburgh, PA, USA

Emmanuel Trélat, CNRS, Laboratoire Jacques-Louis Lions, Sorbonne University, PARIS CEDEX 05, Paris, France

Xu Zhang, School of Mathematics, Sichuan University, Chengdu, Sichuan, China

SpringerBriefs on PDEs and Data Science targets contributions that will impact the understanding of partial differential equations (PDEs), and the emerging research of the mathematical treatment of data science.

The series will accept high-quality original research and survey manuscripts covering a broad range of topics including analytical methods for PDEs, numerical and algorithmic developments, control, optimization, calculus of variations, optimal design, data driven modelling, and machine learning. Submissions addressing relevant contemporary applications such as industrial processes, signal and image processing, mathematical biology, materials science, and computer vision will also be considered.

The series is the continuation of a former editorial cooperation with BCAM, which resulted in the publication of 28 titles as listed here: https://www.springer.com/gp/mathematics/bcam-springerbriefs

Nik Cunniffe • Frédéric Hamelin •
Abderrahman Iggidr • Alain Rapaport •
Gauthier Sallet

Identifiability and Observability in Epidemiological Models

A Primer

 Springer

Nik Cunniffe
Department of Plant Sciences
University of Cambridge
Cambridge, UK

Frédéric Hamelin
Department of Ecology
Institut Agro
Rennes, France

Abderrahman Iggidr
Institut Élie Cartan de Lorraine
INRIA
Metz, France

Alain Rapaport
MISTEA
INRAE
Montpellier, France

Gauthier Sallet
University of Lorraine
Metz, France

ISSN 2731-7595   ISSN 2731-7609   (electronic)
SpringerBriefs on PDEs and Data Science
ISBN 978-981-97-2538-0   ISBN 978-981-97-2539-7   (eBook)
https://doi.org/10.1007/978-981-97-2539-7

This Springer imprint is published by the registered company Springer Nature Singapore Pte Ltd.
The registered company address is: 152 Beach Road, #21-01/04 Gateway East, Singapore 189721, Singapore

If disposing of this product, please recycle the paper.

The World requires at least ten years to understand a new idea, however important or simple it may be.
Ronald Ross (1902 Nobel Prize)

Preface

In Mathematical Epidemiology, many papers have the following structure:

- A model is proposed.
- Some parameters are given, extracted from the literature.
- Remaining unknown parameters are estimated by fitting the model to some observed data.

Fitting is done usually by using an optimization algorithm with the use, for example, of a least square method or a maximum likelihood estimation. To validate the estimation of parameters, one can use noisy synthetic simulated data obtained from the model for given values of the parameters, to check that the algorithm is able to reconstruct from the data the values of these parameters with accuracy.

One objective of this book is to show that this procedure is not always safe and that an examination of the identifiability of parameters is a prerequisite before a numerical determination of parameters. We will review different methods to study identifiability and observability and then consider the problem of numerical identifiability. Our touchstone will be the most famous, but simple, model in Mathematical Epidemiology, the SIR model of Kermack and Mckendrick [73]. This model received renewed attention with the COVID-19 pandemic [106]. Parameter identifiability analysis addresses the problem of which unknown parameters of an ODE model can uniquely be recovered from observed data. We will show that, even for very simple models, identifiability is far from being guaranteed.

The problem of identifiability for epidemiological models is relatively rarely addressed. For instance, a search in the Mathematical Reviews of the American Mathematical Society[1] for 2020 with `epid*` AND `identifiability` gives only 4 papers, while `epidem*` AND `parameter` returns 68 publications. Only a small subset of the later publications addresses the problem of identifiability. In particular, the following publications consider the problem of identifiability

---

[1] https://mathscinet.ams.org/mathscinet.

in epidemiological models: [19, 33, 50, 51, 68, 72, 81, 87, 91, 99, 107, 119–121, 130, 132]. However, the majority of these papers were published elsewhere than in Biomathematics journals. Note that we make a distinction between publications that address directly the parameter estimation problem in epidemiological models (such as in references [13, 20, 21, 27–29, 37, 56, 57, 63, 101, 113, 132] for instance) and works that study explicitly the identifiability property of models. As explained in this book, this is an intrinsic property to be studied prior to determination of parameters values.

The question of observability, i.e. the ability to reconstruct state variables of the model from measurements, is often considered separately from the problem of identifiability. Either model parameters are known, or an identifiability analysis is performed prior to the study of observability. Indeed, the concepts of identifiability and observability are closely related, as we show in this book. However, for certain models, it is possible to reconstruct state variables with observers, while the model is not identifiable. In other situations, we show that considering jointly identifiability and observability with observers can be a way to solve the identifiability problem. This is another illustration of the utility of the concept of observers. This is why we shall dedicate a fair part of this monograph to reviewing the concept of observers and their practical constructions in epidemiology.

This book is aimed at scientists, researchers and graduate students, who use or develop mathematical models for epidemiology, and who are not yet familiar with the concepts of control science (detectability, observability, observers) applied to this field.

Cambridge, UK   Nik Cunniffe
Rennes, France   Frédéric Hamelin
Metz, France   Abderrahman Iggidr
Montpellier, France   Alain Rapaport
Metz, France   Gauthier Sallet
May 2023

Acknowledgements

The authors are deeply grateful to P.-A. Bliman, C. Lobry, J. Harmand, A. Kubik, T. Sari, M. Sofonea, M. Souza and many other colleagues or students, for exchanges and fruitful discussions that gave them the willingness to write this monograph. The authors thank the support of the French National Research Agency ANR (NOCIME project-ANR-23-CE48-0004 and BEEP project ANR-23-CE35-0012).

Contents

Chapter 1
Introduction

Abstract We introduce the basic concepts and present the structure of the book.

1.1 Definitions

The question of parameter identifiability originates from control theory and is related to observability and controllability concepts [110]. The first appearance is in Kalman [71] and is now sixty years old. Identifiability is related to observability: the observability of a model is the ability to reconstruct the state of a system from the observation, while identifiability is the ability to recover unknown parameters from the observation, when the initial condition is know (we give precise definitions below). In the language of dynamical systems with inputs and outputs, which is the standard paradigm in control systems theory, an input-output relation is defined. Typically, models take the form

$$\begin{cases} \dfrac{dx(t)}{dt} = f(x(t), u(t)), \\[2mm] y(t) = h(x(t)), \end{cases}$$

where $u(t)$ and $y(t)$ are respectively "input" and "output" vectors at time t, and the vector $x(t)$ represents the internal variables of the model at time t. The inputs, represented by the function $u(\cdot)$, also called "controls" or "control variables", are considered as known. The outputs, represented by the function $y(\cdot)$, are called "observations" or "measurements" and are also considered as known. For simplicity, we will only consider systems without control (which is a peculiar case where the controls take values in a singleton). When controls are known, with more information, observability/identifiability is sometimes easier. These problems have rarely been considered for uncontrolled systems whereas many methods have been

N. Cunniffe et al., *Identifiability and Observability in Epidemiological Models*,
SpringerBriefs on PDEs and Data Science,
https://doi.org/10.1007/978-981-97-2539-7_1

1

developed for controlled systems. To be more precise, let us consider a dynamical system defined in an open domain $\mathcal{D} \subset \mathbb{R}^n$

$$\begin{cases} \dot{x}(t) = f(x(t)), \ x(0) = x_0 \in \mathcal{D}, \\ \\ y(t) = h(x(t)), \end{cases} \tag{1.1}$$

where we have denoted $\dot{x}(t) = \dfrac{dx(t)}{dt}$. The ordinary differential equation (ODE) $\dot{x} = f(x)$ is the dynamics and x is called the state of the system. To avoid technical details we will assume in all the book that for any initial condition x_0 in $\mathcal{D}$, there exists a unique solution denoted $x(t, x_0)$ such that $x(0, x_0) = x_0$ and

$$\frac{dx(t, x_0)}{dt} = f(x(t, x_0)).$$

Moreover, we will assume that this solution $x(t, x_0)$ is defined for any time $t \geq 0$, and we shall consider a connected subset Ω of $\mathcal{D}$ of non-empty interior that is positively invariant, which means that for any initial condition $x_0 \in \Omega$, the solution $x(t, x_0)$ belongs to this set for any $t \geq 0$. This is often the case with epidemiological models for which the state vector x naturally evolves in a compact and connected invariant set with non-negative vectors. This situation is also often encountered in biological systems. Throughout the manuscript, Ω will then denote the state space.

The output (or "observation") of the system is given by $h(x)$ where h is a differentiable function $h : x \in \Omega \subset \mathbb{R}^n \mapsto h(x) \in Y \subset \mathbb{R}^m$. The set Y is the output space. We will denote by $h(t, x_0)$, or $y(t, x_0)$ the observation at time t for an initial condition x_0.

Definition 1.1 (Observability) The system (1.1) is observable on Ω if for two distinct initial states x_0, x_0' in Ω, there exists a time $t \geq 0$ such that

$$h(x(t, x_0)) \neq h(x(t, x_0')).$$

This is equivalent to state

$$\left\{ h(x(t, x_0)) = h(x(t, x_0')), \ \forall t \geq 0 \right\} \Rightarrow x_0 = x_0'.$$

Two states x_0, x_0' in Ω are called indistinguishable if we have

$$h(x(t, x_0)) = h(x(t, x_0')), \quad t \geq 0.$$

Indistinguishability means that it is impossible to distinguish the evolution of the system, from two distinct initial conditions, by considering only the observation $y(\cdot)$.

Note that the observability property may depend on the choice of the (positively) invariant set Ω. Consider for instance the system

$$\begin{cases} \dot{x}_a = -x_a x_b, \\ \dot{x}_b = x_a x_b, \\ \\ y = x_b. \end{cases}$$

It is not observable for $\Omega = \{x_a \geq 0, \; x_b \geq 0, \; x_a + x_b \leq 1\}$ because all initial conditions of the form $(x_a, 0)$ cannot be distinguished, while it is observable for $\Omega = \{x_a > 0, \; x_b > 0, \; x_a + x_b \leq 1\}$. If two solutions $(x_a(\cdot), x_b(\cdot))$, $(x_a'(\cdot), x_b'(\cdot))$ generate the same output $y(\cdot)$, one should have

$$\begin{cases} y(t) = x_b(t) = x_b'(t), \\ \dot{y}(t) = x_a(t) y(t) = x_a'(t) y(t), \end{cases} \qquad t \geq 0$$

but solutions in Ω are such that $y(t) \neq 0$ at any $t \geq 0$, which implies $x_a(t) = x_a'(t)$ and thus $(x_a(0), x_b(0)) = (x_a'(0), x_b'(0))$.

Sometimes, we shall simply say that "the system is observable", when $\Omega = \mathcal{D}$ or when there is no ambiguity about the choice of Ω.

Now we consider a system depending on a parameter vector $\theta \in \Theta \subset \mathbb{R}^p$ (sometimes simply called the "parameter θ")

$$\begin{cases} \dot{x}(t) = f(x(t), \theta), \; x(0) = x_0, \\ \\ y(t) = h(x(t, \theta)). \end{cases} \qquad (1.2)$$

We denote by $x(t, x_0, \theta)$ the solution of (1.2) for an initial condition x_0. Then, the definition of observability needs to be adapted as follows.

Definition 1.2 (Observability for Parameterized Systems) The system (1.2) is observable on Ω if whatever are $\theta \in \Theta$ and two distinct initial conditions x_0, x_0' in Ω, there exists a time $t \geq 0$ such that

$$h(x(t, x_0, \theta)) \neq h(x(t, x_0', \theta))$$

or equivalently

$$\left\{ h(x(t, x_0, \theta)) = h(x(t, x_0', \theta)), \; \forall t \geq 0 \right\} \Rightarrow x_0 = x_0'.$$

Let us now come to the identifiability property.

Definition 1.3 (Identifiability) Given an initial state x_0 in Ω, system (1.2) is said to be identifiable if for any distinct θ_1, θ_2 in Θ, there exists $t \geq 0$ such that

$$h(x(t, x_0, \theta_1)) \neq h(x(t, x_0, \theta_2)).$$

The concepts of observability and identifiability are similar. Consider the augmented system which consists in adding the parameter θ as part of the (augmented) state vector with a null dynamics:

$$\begin{cases} \dot{x}(t) = f(x(t), \theta), \\ \dot{\theta} = 0, \\ y(t) = h(x(t, \theta)). \end{cases} \tag{1.3}$$

Proposition 1.1 *If the system (1.3) is observable on $\Omega \times \Theta$, then system (1.2) is observable on Ω and identifiable for any $x_0 \in \Omega$.*

Proof If system (1.3) is observable then we have the property: $h(x(t, (x_0, \theta_1))) = h(x(t, (x_0', \theta_2)))$ for all $t \geq 0$ implies $(x_0, \theta_1) = (x_0', \theta_2)$, which implies $x_0 = x_0'$ and $\theta_1 = \theta_2$.   □

The converse is not true as it is illustrated by the following example:

$$\begin{cases} \dot{x}_1 = x_1 + \theta\, x_2 \\ \dot{x}_2 = 0 \\ y = x_1 \end{cases} \tag{1.4}$$

System (1.4) is observable and identifiable.

$$\begin{cases} \dot{x}_1 = x_1 + \theta\, x_2 \\ \dot{x}_2 = 0 \\ \dot{\theta} = 0 \\ y = x_1 \end{cases} \tag{1.5}$$

System (1.5), whose state is (x_1, x_2, θ), is not observable since the output is the same for initial conditions $(x_{1,0}, x_{2,0}, \theta)$ and $(\bar{x}_{1,0}, \bar{x}_{2,0}, \bar{\theta})$ satisfying $x_{1,0} = \bar{x}_{1,0}$ and $x_{2,0}\theta = \bar{x}_{2,0}\bar{\theta}$.

At several places in the following, we shall consider this augmented dynamics.

Actually, for an epidemiological model it is unlikely to know the initial condition and it has long been recognized that initial conditions play a role in identifying the parameters [45, 83, 108, 118, 132].

What we have called identifiability is also known as *structural identifiability*. This expression has been coined by R. Bellman and K.J. Åström [18] in 1970. This is to stress that identifiability depends only on the dynamics and the observation, under ideal conditions of noise-free observations and error-free model. This is a mathematical and *a priori* problem [68].

Sometimes, it happens that the concept of observability is too strong in the sense that a system in the form (1.1) does not satisfy the observability property but it is still possible to know part of the initial condition. A relaxed definition of observability is as follows.

Definition 1.4 (Partial Observability) Let φ be a smooth map from Ω to $\varphi(\Omega) \subset \mathbb{R}^{n'}$. We say that the system (1.1) is partially observable on Ω with respect to φ if for two initial states x_0, x_0' in Ω with $\varphi(x_0) \neq \varphi(x_0')$, there exists a time $t \geq 0$ such that

$$h(x(t, x_0)) \neq h(x(t, x_0')).$$

Two states x_0, x_0' in Ω are called indistinguishable with respect to φ if one has $\varphi(x_0) \neq \varphi(x_0')$ and

$$h(x(t, x_0)) = h(x(t, x_0')), \quad t \geq 0.$$

Alternatively, we shall say the function φ is observable on Ω.

In a similar way, one can relax the definition of identifiability as follows.

Definition 1.5 (Partial Identifiability) Let ϑ be a smooth map from Θ to $\vartheta(\Theta) \subset \mathbb{R}^{p'}$. Given an initial state x_0 in Ω, system (1.2) is said to be identifiable with respect to ϑ if for any θ_1, θ_2 in Θ with $\vartheta(\theta_1) \neq \vartheta(\theta_2)$, there exists $t \geq 0$ such that

$$h(x(t, x_0, \theta_1)) \neq h(x(t, x_0, \theta_2)).$$

Alternatively, we shall say the parameter function ϑ is identifiable.

Example 1.1 Consider the system in $\mathbb{R}^3$

$$\begin{cases} \dot{x}_a = -\alpha x_b \\ \dot{x}_b = -\beta x_c \\ \dot{x}_c = \gamma \end{cases}$$

where α, β, γ are parameters. Solutions of this system can be made explicit:

$$x_a(t) = x_a(0) - \alpha x_b(0)t + \frac{\alpha\beta}{2}x_c(0)t^2 + \frac{\alpha\beta\gamma}{6}t^3,$$

$$x_b(t) = x_b(0) - \beta x_c(0)t - \frac{\beta\gamma}{2}t^2,$$

$$x_c(t) = x_c(0) + \gamma t.$$

Here, we consider $\Omega = \mathcal{D} = \mathbb{R}^3$ and $\Theta = \{(\alpha, \beta, \gamma); \ \alpha > 0, \ \beta > 0, \ \gamma > 0\}$. Let us show that the system is observable for the observation map $h(x_a, x_b, x_c) = x_a$. We consider two initial conditions $(x_a(0), x_b(0), x_c(0))$, $(x'_a(0), x'_b(0), x'_c(0))$ such that their respective solutions verify $x_a(t) = x'_a(t)$ for any $t \geq 0$. From the expression of the solution $x_a(\cdot)$, this implies that one has

$$- \alpha x_b(0)t + \frac{\alpha\beta}{2} x_c(0)t^2 = -\alpha x'_b(0)t + \frac{\alpha\beta}{2} x'_c(0)t^2$$

for any $t \geq 0$, and for $t > 0$ one gets by differentiating with respect to t

$$- \alpha x_b(0) + \frac{\alpha\beta}{2} x_c(0)t = -\alpha x'_b(0) + \frac{\alpha\beta}{2} x'_c(0)t.$$

When $t > 0$ tends to 0, one obtains $x_b(0) = x'_b(0)$ and then $x_c(0) = x'_c(0)$. Therefore, the two initial conditions coincide and there is no indistinguishable distinct initial condition. The system is thus observable.

Consider now the observation map $h(x_a, x_b, x_c) = x_b$. When two initial conditions are such that their solutions verify $x_b(t) = x'_b(t)$, one gets from the expression of the solution $x_b(\cdot)$

$$- \beta x_c(0)t = -\beta x'_c(0)t$$

for any $t > 0$ and thus $x_c(t) = x'_c(t)$. However, having $x_a(0) \neq x'_a(0)$ with $x_b(t) = x'_b(t)$ and $x_c(t) = x'_c(t)$ gives exactly the same solution $x_b(\cdot)$, $x_c(\cdot)$. The system is not observable but it is partially observable with respect to the map

$$\varphi(x_a, x_b, x_c) = (x_b, x_c).$$

Let us now study the identifiability of the system, first for the observation map $h(x_a, x_b, x_c) = x_a$. For a given initial condition with $x_b(0) \neq 0$ and $x_c(0) \neq 0$, if two sets of parameters $(\alpha_1, \beta_1, \gamma_1)$, $(\alpha_2, \beta_2, \gamma_2)$ give the same solution $x_a(\cdot)$, one has

$$- \alpha_1 x_b(0)t + \frac{\alpha_1\beta_1}{2} x_c(0)t^2 + \frac{\alpha_1\beta_1\gamma_1}{6} t^3 = -\alpha_2 x_b(0)t + \frac{\alpha_2\beta_2}{2} x_c(0)t^2 + \frac{\alpha_2\beta_2\gamma_2}{6} t^3,$$

for any $t \geq 0$, and for $t > 0$ one gets

$$- \alpha_1 x_b(0) + \frac{\alpha_1\beta_1}{2} x_c(0)t + \frac{\alpha_1\beta_1\gamma_1}{6} t^2 = -\alpha_2 x_b(0) + \frac{\alpha_2\beta_2}{2} x_c(0)t + \frac{\alpha_2\beta_2\gamma_2}{6} t^2.$$

When $t > 0$ tends to 0, one obtains $\alpha_1 = \alpha_2$ and thus one has

$$\frac{\alpha_1\beta_1}{2} x_c(0)t + \frac{\alpha_1\beta_1\gamma_1}{6} t^2 = \frac{\alpha_2\beta_2}{2} x_c(0)t + \frac{\alpha_2\beta_2\gamma_2}{6} t^2$$

for any $t > 0$. In a similar way, one can show that one has $\alpha_1 \beta_1 = \alpha_2 \beta_2$ and then $\alpha_1 \beta_1 \gamma_1 = \alpha_2 \beta_2 \gamma_2$, which proves that one has $\beta_1 = \beta_2$ and $\gamma_1 = \gamma_2$ (remind that parameters are assumed to be non null). The system is thus identifiable.

When the observation map is $h(x_a, x_b, x_c) = x_b$, having the same output $x_b(\cdot)$ implies

$$- \beta_1 x_c(0) t - \frac{\beta_1 \gamma_1}{2} t^2 = - \beta_2 x_c(0) t - \frac{\beta_2 \gamma_2}{2} t^2$$

for any $t \geq 0$ As before, one obtains $\beta_1 = \beta_2$ and $\gamma_1 = \gamma_2$ but having $\alpha_1 \neq \alpha_2$ with $\beta_1 = \beta_2$ and $\gamma_1 = \gamma_2$ provides the same output $x_b(\cdot)$. Indeed the dynamics of x_b and x_c are decoupled from the one of x_a, where parameter α appears only. The system is not identifiable but is partially identifiable with respect to the map $\vartheta(\alpha, \beta, \gamma) = (\beta, \gamma)$.

1.2 Historical Notes

The observability concept has been introduced by Kalman [71] in the sixties for linear systems. For nonlinear systems, observability has been characterized circa the seventies [58, 62]. The definition is given by Hermann and Krener in the framework of differential geometry. Identifiability and structural identifiability have been introduced in compartmental analysis in 1970 by Bellman and Åström [18] in a paper that appeared in a biomathematics journal. The problem of identifiability is now addressed in text-books [82, 124, 126]. Numerical identifiability of linear control systems is implemented in softwares such as Matlab and Scilab.

Identifiability of nonlinear systems has been addressed in different contexts and the first systematic approach is by Tunali and Tarn in 1987 [118] also in the differential geometry framework. The introduction of the concepts of differential algebra in control theory is due to Fliess around 1990 [45, 46, 52] followed by Glad [53, 83]. Identifiability is a general problem which has received different names depending on the community:

- observation, identification,
- data assimilation,
- inverse problem,
- parameters estimation.

This was accompanied by the development of some specific software to test identifiability (see for instance [19, 33, 65]).

"Data assimilation" is mainly used in meteorology and oceanography [76, 115]. A direct (as opposed to inverse) problem is considering a model which, when introducing an input, gives an observed output. The parameters are considered as known. Conversely, the "inverse problem" is to reconstruct the parameters from the knowledge of the output [116]. Finally, "parameters estimation" is used in the probability and statistics domains [5, 23, 77, 79, 94, 105, 117].

1.3   Identifiability in Mathematical Epidemiology

Identifiability is well known in biomathematics from the seventies, as already mentioned with the paper of Bellman and Åström [18]. However, considering identifiability in mathematical epidemiology is relatively recent [50, 51, 87, 99, 107, 119, 132]. The first paper, to our knowledge, considering identifiability of an intra-host model of HIV is by Xia and Moog [132], and has been published in 2003 in a journal of automatic control.

1.4   The Concept of Observers

The construction of an *observer* is based on an estimation approach different from statistical methods: it consists of determining a dynamical system (called an "observer") whose input is the vector $y(\cdot)$ of measures acquired over time, and whose state is an estimate $\hat{x}(t)$ of the (unknown) true state $x(t)$ of the system at time t.

An observer estimates $x(t)$ continuously over time and without anticipation, in the sense that the estimate $\hat{x}(t)$ is updated at each instant t through its dynamics as measurement $y(t)$ is available, without requiring the knowledge of any future measurement. This is why an observer is sometimes also called a "software sensor". Since the estimate $\hat{x}(t)$ is given by the solution of a system of differential equations, the main idea behind an observer is the paradigm of *integrating instead of differentiating* the signal $y(\cdot)$. Note that although an observer is primarily devoted to state estimation, an observer can also aim to reconstructing simultaneously states and parameters, when some parameters are unknown (in this case a parameter vector p is simply considered a part of the system dynamics with $\dot{p} = 0$).

The most well-known observer is the so-called *Luenberger observer* [84] that is recalled in Chap. 4, and that has inspired most of the existing observers (several ones are discussed in Chap. 4). However, observers are yet relatively unpopular in Mathematical Epidemiology, comparatively to other application domains (such as mechanics, aeronautics, automobile, etc.), apart some few works such as in [1, 6] for instance. The aim of the present review is also to promote the development and use of observers for epidemiological models.

Chapter 4 presents the theoretical background of observers construction and their convergence as estimators based on the model equations, independently of the quality of real data. In a complementary way, Chap. 5 discusses some implementation issues when observers are used with real world data that could be corrupted with noise.

Chapter 2
Mathematical Foundations

Abstract We give mathematical characterizations and properties of observability and identifiability.

2.1 Preliminaries

Here and in all the following chapters, we shall consider that the maps f and h that define the system

$$\begin{cases} \dot{x} = f(x), & x \in \Omega \subset \mathbb{R}^n, \\ y = h(x) \in Y \subset \mathbb{R}^m \end{cases} \tag{2.1}$$

are analytic at any point $x \in \Omega$. We shall denote by $x(t, x_0)$ the solution of $\dot{x} = f(x)$ for the initial condition $x(0) = x_0$.

We recall that a function $\varphi : \mathcal{D} \longrightarrow \mathbb{R}$ is analytic on an open domain $\mathcal{D}$ of $\mathbb{R}^n$ (we will write $\varphi \in C^\omega(\mathcal{D}, \mathbb{R})$) if it is C^∞ (i.e. infinitely differentiable) and its Taylor series converges locally, that is for any x_0 in $\mathcal{D}$ there exists a neighborhood $\mathcal{V}$ of x_0 in $\mathcal{D}$ such that

$$\varphi(x) = \lim_{n \to +\infty} \sum_{k=0}^{n} \frac{\varphi^{(k)}(x_0)}{k!}(x - x_0)^k, \quad x \in \mathcal{V}.$$

Indeed, up to our knowledge, the great majority of epidemiological model in the literature are analytic.

The map $f : \Omega \longrightarrow \mathbb{R}^n$ is also called a *vector field* on Ω, as its image for each $x \in \Omega$ is a (velocity) vector in $\mathbb{R}^n$.

© The Author(s), under exclusive license to Springer Nature Singapore Pte Ltd. 2024
N. Cunniffe et al., *Identifiability and Observability in Epidemiological Models*,
SpringerBriefs on PDEs and Data Science,
https://doi.org/10.1007/978-981-97-2539-7_2

Given a C^∞ function $g : \mathbb{R}^n \longrightarrow \mathbb{R}$, the classical definition of Lie derivative of g with respect to the vector field f is given by

$$\mathcal{L}_f(g)(x) = \frac{d}{dt} g(x(t, x)) \Big|_{t=0} = \langle \nabla g(x) | f(x) \rangle, \quad x \in \Omega \tag{2.2}$$

where ∇g is the gradient of g and $\langle \,|\, \rangle$ the inner product of $\mathbb{R}^n$. For $k > 1$, we define by induction

$$\mathcal{L}_f^k(g)(x) = \mathcal{L}_f(\mathcal{L}_f^{k-1}(g))(x)$$

Note that $\mathcal{L}_f^k(h)(x)$ is the value of the k-th time derivative at time 0 of the output $y(t) = h(x(t, x))$ along the solution of system (2.1) with initial state x, but considered as function of x:

$$\mathcal{L}_f(h)(x) = \dot{y}(x),$$
$$\mathcal{L}_f^2(h)(x) = \ddot{y}(x),$$

$$\vdots$$

$$\mathcal{L}_f^k(h)(x) = y^{(k)}(x)$$

For a C^∞ vector-valued function $g : \mathbb{R}^n \longrightarrow \mathbb{R}^q$, the Lie derivative $\mathcal{L}_f(g)(x)$ at point x is the vector of Lie derivatives of each component g_j $(j = 1 \cdots q)$

$$\mathcal{L}_f(g)(x) = \begin{bmatrix} \mathcal{L}_f(g_1)(x) \\ \vdots \\ \mathcal{L}_f(g_q)(x) \end{bmatrix}$$

When f and g are linear, that is $f(x) = Ax$ and $g(x) = Cx$ for some matrices A and C, then $\mathcal{L}_f^k(g)(x)$ is directly equal to $CA^k x$.

We shall denote $^\top$ the transposition operator for a vector or a matrix.

2.2 Observability

In this section, we introduce the *observation space* and the concept of *local observability* that can be checked with the help of differential calculus in terms of a (local) *rank condition*. For linear systems, this rank condition depends only on the matrices defining the system and the output and ensures the observability property as defined in Sect. 1.1. For analytic nonlinear systems, we shall see that this rank condition depends on the vector field and the output function as well as on the initial condition x_0 and only ensures a "local" observability of the system on a neighborhood of x_0.

The observability definition 1.1 given in Chap. 1 states a global observability property. It also can be formulated as follows.

Definition 2.1 (Observability) The analytic system (2.1) is observable if for any initial conditions x_1 and x_2, for any $T > 0$, one has: $h(x(t, x_1)) = h(x(t, x_2))$ for all $t \in [0, T]$ implies $x_1 = x_2$. This is equivalent to say that the map:

$$\Omega \to C^\omega([0, T], \mathbb{R}^m)$$

$$x_0 \mapsto \left\{ \begin{array}{ccc} [0, T] & \to & \mathbb{R}^m \\ t & \mapsto & h(x(t, x_0)) \end{array} \right\}$$

is injective.

Observability for analytic systems means that the initial state x_0 (and therefore the trajectory starting from this initial state) can be uniquely determined by the knowledge of the data of the output $y(\cdot)$ on any nontrivial time interval.

The components of the observation map h are denoted by $h = (h_1, \cdots, h_m)$. Each h_i is a C^∞ function from the state space $\mathbb{R}^n$ to $\mathbb{R}$.

Definition 2.2 ([62]) The observation space O of (2.1) is the subspace of the vector space $C^\infty(\mathbb{R}^n, \mathbb{R})$ containing h_i and invariant by the Lie derivative $\mathcal{L}_f$.

The observation space is generated by the different Lie derivatives of the h_i:

$$O = \text{span}_{\mathbb{R}} \left\{ \mathcal{L}_f^k h_i : i = 1, \ldots, m, \ k \in \mathbb{N} \right\}$$
$$= \text{span}_{\mathbb{R}} \left\{ h_i, \mathcal{L}_f h_i, \mathcal{L}_f^2 h_i, \ldots : i = 1, \ldots, m \right\}.$$

The observation space O contains the observation function (also called the output function) and all the derivatives of the output function along the system trajectories. For a linear system $f(x) = Ax$ and $y = h(x) = Cx$, the observation space is generated by the n functions

$$Cx, \ CAx, \ \ldots, \ CA^{n-1}x.$$

We recall that, thanks to Cayley-Hamilton Theorem, A^k for $k \geq n$ is a linear combination of A^q with $q \leq n - 1$.

We have the following result that relates the observation space to the observability property.

Theorem 2.1 *For an analytic system (i.e., f and h are analytic functions) the observability is equivalent to the separation of the points of the state space $\mathbb{R}^n$ by O i.e., if $x_1 \neq x_2$ there exists $g \in O$ such that $g(x_1) \neq g(x_2)$.*

Proof By analyticity we have

$$y(t, x_0) = h(x(t, x_0)) = \sum_{k \geq 0} \left(\frac{d^k}{ds^k} h(x(s, x_0)) \bigg|_{s=0} \right) \frac{t^k}{k!},$$

but, by induction we have the following relation

$$\frac{d^k}{ds^k} h(x(s, x_0)) \bigg|_{s=0} = \mathcal{L}_f^k h(x_0).$$

Then a necessary and sufficient condition to distinguish $x_1 \neq x_2$ is that there exists k such that

$$\mathcal{L}_f^k h (x_1) \neq \mathcal{L}_f^k h (x_2).$$

Example 2.1 (Application to a Virus Dynamics Model) We consider a simple model of an HIV-1 infection [100]:

$$
\begin{cases}
\dot{T} = \Lambda - \mu_T T - \beta V T, \\
\dot{T}^* = \beta V T - \delta T^*, \\
\dot{V} = r \delta T^* - c V, \\
\\
y = V,
\end{cases}
\tag{2.3}
$$

where T, T^*, V denote the concentrations of uninfected (healthy) and infected host cells, and freevirions, respectively. The rate of infection is given by $\beta V T$, with β being the infection rate constant. The parameters δ and c are the removal rates of the infected cells and virus particles respectively. The healthy cells (T) are produced at a rate Λ and μ_T is the death rate per T cell. It is assumed that on average each productively infected cell produces r (a positive integer) virions during its lifetime, so the per-capita rate of viral production for an infected cell is given by $r \delta$. All parameters are positive. We assume that the measurement of the viral load is available.

Let us show that system (2.3) is observable on the set $\{(T, T^*, V) \in \mathbb{R}^3 : T \geq 0, T^* \geq 0, V > 0\}$. We shall prove that the observation space associated to System (2.3) separates the points of set $\{(T, T^*, V)^\top \in \mathbb{R}^3 : T \geq 0, T^* \geq 0, V > 0\}$.

Here we have $x = (T, T^*, V)^\top$ and

$$
f(x) = \begin{bmatrix} \Lambda - \mu_T T - \beta V T \\ \beta V T - \delta T^* \\ r \delta T^* - c V \end{bmatrix}
$$

and the measurable output $y = h(x) = V$. Computing the Lie-derivatives of the output gives

$$\mathcal{L}_f h(x) = \langle \nabla h(x) | f(x) \rangle = \langle \begin{bmatrix} 0 \\ 0 \\ 1 \end{bmatrix} \Big| \begin{bmatrix} \Lambda - \mu_T\, T - \beta\, V\, T \\ \beta\, V\, T - \delta\, T^* \\ r\, \delta\, T^* - c\, V \end{bmatrix} \rangle = r\, \delta\, T^* - c\, V$$

and

$$\mathcal{L}_f^2 h(x) = \langle \nabla \mathcal{L}_f h(x) | f(x) \rangle = \langle \begin{bmatrix} 0 \\ r\,\delta \\ -c \end{bmatrix} \Big| \begin{bmatrix} \Lambda - \mu_T\, T - \beta\, V\, T \\ \beta\, V\, T - \delta\, T^* \\ r\, \delta\, T^* - c\, V \end{bmatrix} \rangle$$

$$= r\, \delta\, \beta\, T\, V - r\, \delta\, (c + \delta)\, T^* + c^2\, V.$$

The explicit expression of $\mathcal{L}_f^3 h(x)$ is quite long but it can be easily shown that $\mathcal{L}_f^3 h(x)$ can be expressed as a function of $h(x)$, $\mathcal{L}_f h(x)$, $\mathcal{L}_f^2 h(x)$ and the parameters of the system. So to study observability of System (2.3), one does not need to compute Lie-derivative of order higher than 2.

The observation space O contains the functions

$$g_1(x) = V,$$

$$g_2(x) = r\, \delta\, T^* - c\, V,$$

$$g_3(x) = r\, \delta\, \beta\, T\, V - r\, \delta\, (c + \delta)\, T^* + c^2\, V.$$

It is easy to show

$$\left\{ g_1(x) = g_1(\bar{x}),\ g_2(x) = g_2(\bar{x}),\ g_3(x) = g_3(\bar{x}) \right\} \implies V = \bar{V},\ T^* = \bar{T}^*$$

and $T = \bar{T}$ if $V \neq 0$. Hence, these functions separate the points of $\mathbb{R}_+^3 \setminus \{(T, T^*, 0)\}$ but they do not separate points of the form $(T, T^*, 0)$ and $(\bar{T}, T^*, 0)$ with $\bar{T} \neq T$. Thus System (2.3) is observable on the set $\{(T, T^*, V) \in \mathbb{R}^3 : T \geq 0,\ T^* \geq 0,\ V > 0\}$.

Applying Theorem 2.1 to linear systems

$$\begin{cases} \dot{x} = Ax, & x \in \mathbb{R}^n, \\ y = Cx, & y \in \mathbb{R}^m, \end{cases} \tag{Σ_L}$$

allows to obtain a simple algebraic necessary and sufficient condition for observability.

Proposition 2.1 *The linear system* (Σ_L) *is observable if and only if the* observability matrix

$$O_{(C,A)} = \begin{bmatrix} C \\ CA \\ \vdots \\ CA^{n-1} \end{bmatrix} \tag{2.4}$$

is of full rank, i.e., rank $O_{(C,A)} = n$.

Indeed, using Theorem 2.1, we have the following successive equivalences:

The linear system (Σ_L) is not observable

$\Updownarrow$

Its observation space O_{Σ_L} does not separate the points of the state space $\mathbb{R}^n$

$\Updownarrow$

$$\exists x_1 \neq x_2 : \forall g \in O_{\Sigma_L}, g(x_1) = g(x_2)$$

$\Updownarrow$

$$\exists x_1 \neq x_2 : \forall i \in \{0, 1, \ldots n-1\}, CA^i x_1 = CA^i x_2$$

$\Updownarrow$

$$\exists x_1 \neq x_2 : x_1 - x_2 \in \ker CA^i, \forall i \in \{0, 1, \ldots n-1\}$$

$\Updownarrow$

$$x_1 - x_2 \in \ker C \cap \ker CA \cap \ker CA^2 \cap \ldots \cap \ker CA^{n-1}$$

$\Updownarrow$

$$\ker O_{(C,A)} \neq \{0\}$$

$\Updownarrow$

$$rank\ O_{(C,A)} < n.$$

The result of Theorem 2.1 can be also reformulated as follows (For the sake of writing simplicity, we consider real scalar output. Extension to vector output is straightforward). Let us define, for $k \in \mathbb{N}$, the map: $x \mapsto H_k(x) = (h(x), \mathcal{L}_f h(x), \ldots, \mathcal{L}_f^k h(x))^\top$. Then we have the following characterization of observability :

Proposition 2.2 ([67]) *Suppose System* (2.1) *is analytic. Then it is observable if and only if* $H_k(x_1) = H_k(x_2)$, *for all* $k \in \mathbb{N}$, *implies that* $x_1 = x_2$.

This is equivalent to say that the following real analytic mapping (from the state space to an infinite dimensional space)

$$x \mapsto H_\infty(x) = (h(x), \mathcal{L}_f h(x), \ldots, \mathcal{L}_f^k h(x), \ldots)^\top$$

is injective.

Remark 2.1 In general there is no value of k to stop. This is illustrated by the following example [67]:

$$\begin{cases} \dot{x} = -x, \quad x \in \mathbb{R} \\ \\ y = h(x) = x \prod_{i=1}^{\infty} (1 - e^{x^2 - i^2})^i \end{cases} \tag{2.5}$$

It has been proved in [67] that

- h is analytic on $\mathbb{R}$,
- system (2.5) is observable,
- $\forall k \in \mathbb{N}^*$, the equation $H_k(x) = H_k(0)$ has countably infinite solutions $x = 0$ and $x = \pm(k+i)$, $i = 0, 1, 2, \dots$

Observability can also be checked using the following proposition.

Proposition 2.3 *If any state* $x \in \Omega$ *can be expressed as a function of the observation* y *and its time derivatives* $y^{(k)} = \mathcal{L}_f^k h(x)$, *that is there exists a map* ϕ *such that* $x = \phi \left(h(x), , \mathcal{L}_f h(x), \dots, \mathcal{L}_f^k h(x), \dots \right)$, *then System (2.1) is observable on* Ω.

Proof Let $x_1 \in \Omega$ and $x_2 \in \Omega$ be such that $H_\infty(x_1) = H_\infty(x_2)$. Then, applying ϕ, we obtain $x_1 = x_2$ which proves that map H_∞ is injective on Ω.   □

Example 2.2 Consider the SIR model of Kermack-McKendrick [73]

$$\begin{cases} \dot{S} = -\beta \dfrac{S}{N} I, \\ \dot{I} = \beta \dfrac{S}{N} I - \gamma I, \\ \dot{R} = \gamma I \end{cases} \tag{2.6}$$

for which the parameters β, γ and N are assumed to be known, where the total population $N = S + I + R$ is constant since $\dot{N} = \dot{S} + \dot{I} + \dot{R} = 0$. We assume that the recovery $y = \gamma I$ is observed. Here we have

$$x = \begin{pmatrix} S \\ I \\ R \end{pmatrix}, \quad f(x) = \begin{pmatrix} -\beta \dfrac{S}{N} I \\ \beta \dfrac{S}{N} I - \gamma I \\ \gamma I \end{pmatrix}, \quad h(x) = \gamma I$$

Clearly, the solutions of the system (2.6) evolve in the positively invariant set (which makes biological sense)

$$\Omega = \{(S, I, R) \mid S > 0, \ I > 0, \ R > 0, \ S + I + R = N\}$$

on which one has $y \neq 0$. Then we have the relations (we can divide by y)

$$S = \frac{N}{\beta} \frac{\dot{y} + \gamma y}{y}, \quad I = \frac{y}{\gamma}, \quad R = N - \frac{N}{\beta} \frac{\dot{y} + \gamma y}{y} - \frac{y}{\gamma}$$

Therefore, the state vector x can be expressed as a function of y and $\dot{y}$. The system is thus observable in Ω. This model is more thoroughly studied in Chap. 3.

The observability (equivalent) conditions given by Theorem 2.1 and Proposition 2.2 are hard to test, since they involve an infinite number of analytic equations and therefore checking the condition that O separates points can be quite a formidable task. This is one of the reasons for which different concepts of local observability have been introduced. One of them is the local weak observability [62]. As we will see, the advantage of the local weak observability as compared to other types of observability is that it is easy to indicate for it simple sufficient conditions of an algebraic nature.

Definition 2.3

– The system (2.1) is said locally observable if, for any x_0, for any open set U containing x_0, x_0 is distinguishable from all the points of U for the restricted system on U.
– The system (2.1) is weakly observable at x if there exists an open neighborhood U of x such that the only point in U which is indistinguishable from x is x itself. The system (2.1) is weakly observable if it is weakly observable at every x.
– The system (2.1)is locally weakly observable if for any x_0 there exists an open set U containing x_0, such that for any neighborhood V with $x_0 \in V \subset U$, x_0 is distinguishable for the system restricted to V from all the points of V.

Intuitively, a system is locally weakly observable if one can instantaneously distinguish each point from its neighbors. The local weak observability can be characterized as follows.

Definition 2.4 ([112]) Let O be the observation space of system (2.1), we define

$$dO = \{d\psi \mid \psi \in O\}$$

where $d\psi(x)$ is the differential of ψ at x.

Definition 2.5 The system (2.1) is said to satisfy the observability rank condition (ORC) at x if the dimension of dO at x satisfies

$$\dim(d\,O(x)) = n$$

where dO is generated by the gradients of the $\mathcal{L}_f^k h$.

Theorem 2.2 (Hermann-Krener [62]) *If the **analytic** system* (2.1) *satisfies the observability rank condition (ORC) at x_0 then* (2.1) *is locally weakly observable at x_0.*

Proof Since $\dim(d\,O(x_0)) = n$, there exist n functions $\varphi_1, \cdots, \varphi_n \in O$ such that the gradients $d\varphi_1(x_0), \cdots, d\varphi_n(x_0)$ are linearly independent. Therefore the function $\Phi : x \mapsto (\varphi_1(x), \cdots, \varphi_n(x))$ has a non-singular Jacobian in x_0. As a consequence, from the Inverse Function Theorem, there exists an open set U containing x_0 where Φ is a bijection.

On any open set $V \subset U$ suppose that we have $h(x(t, x_0)) = h(x(t, x_1))$ for x_0, x_1 in V and $t \in [0, T]$. Then, from the fact that f and h are analytic and

$$\left.\frac{d^k}{dt^k} h(x(t, x_0))\right|_{t=0} = \mathcal{L}_f^k h(x_0),$$

we have

$$h(x(t, x_0)) - h(x(t, x_1)) = \sum_{k \geq 0} \frac{t^k}{k!} \left(\mathcal{L}_f^k h(x_0) - \mathcal{L}_f^k h(x_1) \right) = 0, \text{ for } t \in [0, T].$$

This implies $\mathcal{L}_f^k h(x_0) - \mathcal{L}_f^k h(x_1) = 0$ for all $k \geq 0$ which implies that $\varphi_i(x_0) = \varphi_i(x_1)$ for all $i = 1, \ldots, n$ since each φ_i is a linear combination of the $\mathcal{L}_f^k h$. Therefore $\Phi(x_0) = \Phi(x_1)$ and hence $x_1 = x_0$ since Φ is a bijection. This proves that x_0 is distinguishable from all points of V and hence the analytic system (2.1) is locally weakly observable at x_0. $\qquad\square$

If the observability rank condition is satisfied everywhere the system is locally weakly observable.

A converse result has been proved in [62]:

Proposition 2.4 ([30, 62]) *If the system is locally weakly observable then the rank condition is satisfied almost everywhere, i.e., in an open dense subset of the state space.*

To summarize, we have for analytic systems:

Proposition 2.5 ([30, 62]) *For the **analytic** system* (2.1), *the relationships between the various observability concepts are given in the following diagram*

ORC satisfied almost everywhere

$\Updownarrow$

(2.1) locally observable $\Longrightarrow$ *(2.1) locally weakly observable*

$\Downarrow$ $\qquad\qquad\qquad\qquad\qquad\quad$ $\Updownarrow$

(2.1) observable $\quad\Longrightarrow\quad$ *(2.1) weakly observable*

Remark 2.2 For linear systems, the five properties are equivalent.

Remark 2.3 The analytic system $\dot{x} = 0$, $y = x^3$ is observable on $\mathbb{R}$ and locally weakly observable but the ORC is not satisfied at $x = 0$. However it is satisfied on $\mathbb{R} \setminus \{0\}$ and so it is satisfied almost everywhere.

Remark 2.4 If an analytic system is observable then the observability rank condition (ORC) is satisfied almost everywhere. It must be noticed that the converse is not true: the analytic system

$$\begin{cases} \dot{x} = 1, \ x \in \mathbb{R} \\ \\ y = (\sin x, \cos x) \in \mathbb{R}^2 \end{cases}$$

satisfies the ORC at any $x \in \mathbb{R}$ but it is not observable because the states x and $x + 2k\pi$ are indistinguishable.

It must be emphasized that the study of the observability of analytic systems requires either to deal with an infinite number of analytic equations or to compute the dimension of the linear space generated by the gradients of all Lie derivatives of the output. In general, there is no bound on the number of Lie derivatives necessary to conclude, as seen in example (2.5).

Remark 2.5 For linear systems $\dot{x} = Ax$, $y = Cx$, all the definitions of observability are equivalent to having the *observability matrix*

$$O_{(C,A)} = \begin{bmatrix} C \\ CA \\ \vdots \\ CA^{n-1} \end{bmatrix}$$

of full rank (see for instance [69]).

The observability analysis can also be a way to choose the right measurements, as illustrated on the following example.

Example 2.3 Consider a population model structured in five age classes, whose population sizes are

x_1 for juveniles,
x_2 for subadults capable of reproduction when adults,
x_3 for subadults not capable of reproduction when adults,
x_4 for adults capable of reproduction,
x_5 for adults not capable of reproduction,

and the dynamics is

$$
\begin{cases}
\dot{x}_1 = -\alpha x_1 + \beta x_4 \\
\dot{x}_2 = \frac{\alpha}{2} x_1 - \alpha x_2 - m_1 x_2 \\
\dot{x}_3 = \frac{\alpha}{2} x_1 - \alpha x_3 - m_1 x_3 \\
\dot{x}_4 = \alpha x_2 - m_2 x_4 \\
\dot{x}_5 = \alpha x_3 - m_2 x_5
\end{cases}
$$

(where α is an aging rate, m_1, m_2 are mortality rates, and β is a fecundity rate). If only one sub-population x_i can be targeted for measurement, one can easily check that the only possibility for the system to be observable is to measure the variable $y = x_5$. This system is of the form $\dot{x} = Ax$, $y = Cx$ with

$$
A = \begin{bmatrix}
\overbrace{\begin{matrix} -\alpha & 0 & 0 & \beta \\ \alpha/2 & -\alpha - m_1 & 0 & 0 \\ \alpha/2 & 0 & -\alpha - m_1 & 0 \\ 0 & \alpha & 0 & -m_2 \end{matrix}}^{A_4} & \begin{matrix} 0 \\ 0 \\ 0 \\ 0 \end{matrix} \\
\begin{matrix} 0 & \quad 0 & \quad \alpha & \quad 0 \end{matrix} & -m_2
\end{bmatrix}
$$

that is of the form

$$
\dot{x} = \begin{bmatrix} A_4 & \begin{matrix} x_1 \\ x_2 \\ x_3 \\ x_4 \end{matrix} \\ \alpha x_3 - m_2 x_5 \end{bmatrix}, \quad y = Cx
$$

From this, we see that for any integer $k \geq 1$, A^k is of the form

$$
\begin{bmatrix}
A_4^k & \begin{matrix} 0 \\ 0 \\ 0 \\ 0 \end{matrix} \\
\begin{matrix} * & * & * & * \end{matrix} & (-m_2)^k
\end{bmatrix}.
$$

Therefore, if $C = [c_1\ c_2\ c_3\ c_4\ 0]$ then $CA^k = (*\ *\ *\ *\ 0)$ and hence the observability matrix will be of rank ≤ 4 and so the system is not observable if the output does not depend on x_5.

Now, if $C = [0\ 0\ 0\ 0\ 1]$ then the observability matrix defined by (2.4) is given by:

$$
O_{(C,A)} =
\begin{bmatrix}
0 & 0 & 0 & 0 & 1 \\
0 & 0 & \alpha & 0 & -m_2 \\
\frac{\alpha^2}{2} & 0 & -\alpha(m_1 + m_2 + \alpha) & 0 & m_2^2 \\
-\frac{\alpha^2(2\alpha+m_1+m_2)}{2} & 0 & h_{43} & \frac{\beta\alpha^2}{2} & -m_2^3 \\
h_{51} & \frac{\alpha^3\beta}{2} & h_{53} & -\frac{\beta\alpha^2(2\alpha+m_1+2m_2)}{2} & m_2^4
\end{bmatrix}
$$

with

$$
h_{43} = \left(\alpha^2 + (2m_1 + m_2)\alpha + m_1^2 + m_1 m_2 + m_2^2\right)\alpha,
$$

$$
h_{51} = \left(\frac{3\alpha^2}{2} + \frac{(3m_1 + 2m_2)\alpha}{2} + \frac{m_1^2}{2} + \frac{m_1 m_2}{2} + \frac{m_2^2}{2}\right)\alpha^2,
$$

$$
h_{53} = -\alpha(m_1 + m_2 + \alpha)\left(\alpha^2 + 2\alpha m_1 + m_1^2 + m_2^2\right).
$$

One has $det\ O_{(C,A)} = -\dfrac{\alpha^8\beta^2}{8} \neq 0$. Hence $O_{(C,A)}$ is of full rank which proves the observability of the system when $y = x_5$.

2.3 About Identifiability

Since very often the initial conditions are not known, or partially known, we will consider in the following the problem of joined identifiability and observability, considering the augmented system (1.3). Note that identifiability-only problems are a special case in which $y = x$ (this is why we consider here the more general case of joined identifiability and observability).

Consider a parametrized system

$$
\begin{cases}
\dot{x}(t) = f(x(t), \theta), \ x(0) = x_0, \\
\\
y(t) = h(x(t), \theta),
\end{cases}
\tag{2.7}
$$

with $x \in \mathbb{R}^n$, $y \in \mathbb{R}^m$ and $\theta \in \Theta \subset \mathbb{R}^p$ and assume that system (2.7) is observable for any known value of the parameter $\theta \in \Theta$. Let us denote the observability map parameterized by θ

$$
H_k(x, \theta) =
\begin{bmatrix}
h(x, \theta) \\
\mathcal{L}_f(h)(x, \theta) \\
\vdots \\
\mathcal{L}_f^{k-1}(h)(x, \theta)
\end{bmatrix}
$$

When there exist an integer $k > 1$ and a map $\Phi_\theta : \mathbb{R}^n \mapsto H_k(\mathbb{R}^n, \theta)$ parameterized by θ, such that the output differential equation

$$y^{(k)}(t) = \Phi_\theta(y(t), \dot{y}(t), \cdots, y^{(k-1)}(t)), \quad t \geq 0$$

as a unique solution in $H_k(\mathbb{R}^n, \theta)$, for any $\theta \in \Theta$, one may study the dependency of the map Φ_θ with respect to θ to ensure the identifiability of the system. In particular, when f and h are polynomials with coefficients parameterized by θ, the map Φ_θ is a polynomial with coefficients $c_i(\theta)$ where $c : \Theta \mapsto \mathbb{R}^\nu$ for some ν. The injectivity of the map c is clearly a necessary condition to have identifiability, but this is not sufficient as one can see on the following example.

Example 2.4 Consider the system in $\mathbb{R}_+^2$

$$\begin{cases} \dot{x}_1 = -\theta_1 x_1 + \theta_2 x_2 \\ \dot{x}_2 = -\theta_2 x_2 \\ \\ y = x_1. \end{cases} \tag{2.8}$$

where $\theta \in \Theta = \mathbb{R} \times \mathbb{R}^\star$ (where $\mathbb{R}^\star$ denotes $\mathbb{R} \setminus \{0\}$) is the unknown vector of parameters. One has

$$H_1(x, \theta) = \begin{bmatrix} x_1 \\ -\theta_1 x_1 + \theta_2 x_2 \end{bmatrix}$$

which is invertible w.r.t. x on $\mathbb{R}_+^2$ for any $\theta \in \Theta$

$$H_1^{-1}(z, \theta) = \begin{bmatrix} z_1 \\ \dfrac{z_2 + \theta_1 z_1}{\theta_2} \end{bmatrix}$$

The system is thus observable for any $\theta \in \Theta$. Moreover, one has

$$\ddot{y} = -\theta_1 \dot{y} - \theta_1^2 x_2 = -\theta_2 \dot{y} - \theta_1 (\dot{y} + \theta_2 y)$$

that is

$$\ddot{y} = \Phi_\theta(y, y') := -(\theta_1 + \theta_2) \dot{y} - \theta_1 \theta_2 y$$

Clearly the application $\theta \mapsto (\theta_1 + \theta_2, \theta_1 \theta_2)$ is injective on Θ, but the system is not identifiable if $x_2(0) = 0$: the solution verifies $x_2(t) = 0$ for any $t > 0$ and x_1 is solution of $\dot{x}_1 = -\theta_1 x_1$ independently of the value of θ_2.

A natural (and usual) way to ensure the joined observability-identifiability property is to consider the augmented state:

$$\tilde{x} = \begin{bmatrix} x \\ \theta \end{bmatrix} \in \mathbb{R}^n \times \Theta$$

and require the observability of the extended dynamics

$$\begin{cases} \dfrac{d}{dt}\tilde{x}(t) = \tilde{f}(\tilde{x}(t)) := f(x(t), \theta), \quad \tilde{x}(0) = (x(0), \theta) \\[2mm] y(t) = \tilde{h}(\tilde{x}(t))) := h(x(t), \theta) \end{cases}$$

and we fall back on a problem of pure observability.

Remark 2.6 A theoretical answer to the problem of parameters reconstruction when measuring the whole state x in $\mathbb{R}^n$ has been given by D. Aeyels [3, 4] and E. Sontag [111] in different form. For an (analytic) system with r parameters, it is *generically* sufficient to choose $2r + 1$ measures at different times to distinguish two different states (the term *generically* means here that for any system except for a non dense subset of systems among all the analytic systems in $\mathbb{R}^n$).

2.4 Identifiability Does Not Necessarily Require Observability

Let us stress that identifiability does not necessarily imply observability. It can happen that the knowledge of the output function $y(\cdot)$ allows to reconstruct uniquely the set of parameters, but not necessarily the state variables of the system. We give below an example of such a situation.

Example 2.5 The following "academic" model is identifiable but not observable.

$$\begin{cases} \dot{x}_1 = -\dfrac{\alpha}{2}(x_1 + x_2) \\[2mm] \dot{x}_2 = \dfrac{\alpha}{2}(x_1 - x_2) \\[4mm] y = h(x_1, x_2) = \dfrac{1}{2}(x_1^2 + x_2^2) \end{cases} \qquad (2.9)$$

One immediately gets

$$\dot{y} = -\alpha\, y.$$

For (unknown) initial conditions $(x_1(0), x_2(0)) \neq (0, 0)$, one has $y(0) > 0$ and y is thus a positive function. Then one obtains $\alpha = -\dot{y}/y$: the system is identifiable on $\mathbb{R}^2 \setminus \{0\}$.

Compute now further derivatives: $\ddot{y} = \alpha^2 y, \cdots, y^{(p)} = (-1)^p \alpha^p y$. Formally, one gets

$$
\text{Jac } [h, \mathcal{L}_f h, \mathcal{L}_f^2 h, \cdots, \mathcal{L}_f^p h] = \begin{bmatrix} x_1 & -\alpha x_1 & \alpha^2 x_1 & \cdots & (-1)^p \alpha^p x_1 \\ x_2 & -\alpha x_2 & \alpha^2 x_2 & \cdots & (-1)^p \alpha^p x_2 \\ 0 & -y & 2\alpha y & \cdots & \alpha(-1)^p \alpha^{p-1} y \end{bmatrix},
$$

which is of rank 2 for any $(x_1, x_2) \in \mathbb{R}^2 \setminus \{0\}$ and for any positive integer p. The parameter α is identifiable, but the system is not observable. Consider another solution $(\xi_1(\cdot), \xi_2(\cdot))$ for the initial condition $\xi_1(0) = -x_1(0), \xi_2(0) = -x_2(0)$. One can straightforwardly check that this solution verifies $\xi_1(t) = -x_1(t), \xi_2(t) = -x_2(t)$ with the same output $y(t)$ for any $t \geq 0$. Therefore, these two solutions cannot be distinguished, which shows that the system is not observable.

Another way to show the non-observability is to remark that we have

$$
y(t) = e^{-\alpha t} y(0) = \frac{1}{2} e^{-\alpha t} \left(x_1(0)^2 + x_2(0)^2 \right) = \frac{1}{2} e^{-\alpha t} \left(\xi_1(0)^2 + \xi_2(0)^2 \right)
$$

for any $(\xi_1(0), \xi_2(0)) \in S_0$ where S_0 is the circle centered at the origin with radius $r_0 = \sqrt{x_1(0)^2 + x_2(0)^2}$. Thus the output cannot distinguish the solutions emanating from different points of this circle. Hence System (2.9) is non observable.

This can also be proved by remarking that one has

$$
\mathcal{L}_f^p h(x) = (-1)^p \alpha^p h(x) = \frac{(-1)^p \alpha^p}{2} (x_1^2 + x_2^2).
$$

This means that the associated observation space is generated by the function $x_1^2 + x_2^2$ that does not separate the points of $\mathbb{R}^2$ and hence, according to Theorem (2.1), system (2.9) is non observable.

2.5 Identifiability via Decoupled Variables

Sometimes it is difficult to prove identifiability of a model in the original set of coordinates. Let us show the interest of considering other variables that possess "good" properties. We consider a system in $\mathbb{R}^n$ parameterized by $\theta \in \Theta \subset \mathbb{R}^p$ of the following form

$$
\begin{cases} \dot{x} = f(x, \theta), & x(0) = x_0 \in X \\ \\ y = h(x) \end{cases}
\tag{2.10}
$$

where $X \subset \mathbb{R}^n$ is positively invariant for any $\theta \in \Theta$.

Proposition 2.6 ([22]) *Assume that the following properties hold.*

1. The map f verifies

$$\forall \theta_1, \theta_2 \in \Theta, \ \forall x \in X, \quad \theta_1 \neq \theta_2 \Longrightarrow f(x, \theta_1) \neq f(x, \theta_2). \tag{2.11}$$

2. There exist smooth maps g, $\tilde{g}$ and l such that

   a. for any solution $x(\cdot)$ of (2.10) in X, $w(t) := g(x(t), y(t)) \in W \subset \mathbb{R}^m$ verifies

$$\dot{w}(t) = l(w(t), y(t)), \quad t \geq 0 \tag{2.12}$$

   b. for any $x \in X$, one has

$$w = g(x, h(x))) \Longleftrightarrow x = \tilde{g}(w, h(x)) \tag{2.13}$$

Then the system (1.1) is identifiable over Θ for any initial condition in X.

Proof Fix $x_0 \in X$ and denote by $x_\theta(\cdot)$ the solution of (2.10) for the parameter $\theta \in \Theta$. Consider θ_1, θ_2 in Θ that give the same output function: $h(x_{\theta_1}(t)) = h(x_{\theta_2}(t)) = y(t)$ for any $t \geq 0$.

Let $w(\cdot)$ be the (unique) solution of the Cauchy problem

$$\dot{w} = l(w, y(t)), \quad w(0) = g(x_0, h(x_0)).$$

Then, one has

$$w(t) = g(x_{\theta_i}(t), y(t)) \quad t \geq 0, \quad i = 1, 2,$$

and from property (2.13), one obtains

$$\tilde{g}(w(t), y(t)) = x_{\theta_i}(t) \quad t \geq 0, \quad i = 1, 2,$$

that is $x_{\theta_1}(t) = x_{\theta_2}(t)$ for any $t > 0$. Therefore, for any $t > 0$ one has also $\dot{x}_{\theta_1}(t) = \dot{x}_{\theta_2}(t)$ or $f(x(t), \theta_1) = f(x(t), \theta_2)$ where $x(t) = x_{\theta_1}(t) = x_{\theta_2}(t)$. Finally, from condition (2.11), we deduce that one has necessarily $\theta_1 = \theta_2$, which shows the identifiability of the system. $\qquad \square$

This result states that when a system is identifiable when measuring the whole state x (condition (2.11)), and there exits a variable w whose dynamics is *decoupled* in the sense that it depends on w and y only (condition (2.12)) such that the map $(x, y) \mapsto w$ is invertible with respect to x (condition (2.13)), then the system is identifiable when measuring y only. Up to our knowledge, this approach has not been deployed in the literature. Let us illustrate this result on an intra-host model for malaria infection [20].

Example 2.6 (Malaria Model) The state vector is $x = (S, I_1, \ldots, I_5, M)^\top$ in $\mathbb{R}_+^7$, where S is the concentration of uninfected erythrocytes in the blood, I_i are the concentrations of infected erythrocytes in different age classes, and M is the concentration of free merozoites. The dynamics is given by the following system

$$
\begin{cases}
\dot{S} = \Lambda - \mu_S S - \beta SM, \\
\dot{I}_1 = \beta SM - (\gamma_1 + \mu_1) I_1, \\
\dot{I}_2 = \gamma_1 I_1 - (\gamma_2 + \mu_2) I_2, \\
\quad \vdots \\
\dot{I}_5 = \gamma_4 I_4 - (\gamma_5 + \mu_5) I_5, \\
\dot{M} = r \gamma_5 I_5 - \mu_M M - \beta SM,
\end{cases}
\tag{2.14}
$$

where the different parameters are

Λ: recruitment of the healthy red blood cells (RBC).
β: rate of infection of RBC by merozoites.
μ_S: natural death rate of healthy cells.
μ_i: natural death rate of i-th stage of infected cells.
γ_i: transition rate from i-th stage to $(i + 1)$-th stage of infected cells.
r : number of merozoites released by the late stage of infected cells.
μ_M : natural death rate of merozoites.

The two first stages of infected erythrocytes (I_1 and I_2) correspond to the concentration of free circulating parasitized erythrocytes than can be observed (seen on peripheral blood smears). Typically, the quantity

$$
y(t) = h(x(t)) = I_1(t) + I_2(t)
$$

is measured at any time t. Among parameters in (2.14), most of them (μ_i, γ_i, and r) are known or at least widely accepted by the community, but the infection rate β, which is playing a crucial role, is unknown and cannot be estimated by biological considerations. Let us then write the dynamics (2.14) as $\dot{x} = f(x, \beta)$. It takes the form

$$
\begin{cases}
\dot{x} = f(x, \beta) := A x + \beta SM \, E + \Lambda e_1, \\
y = C x
\end{cases}
\tag{2.15}
$$

with

$$A = \begin{bmatrix} -\mu_S & 0 & 0 & 0 & 0 & 0 & 0 \\ 0 & -\gamma_1 - \mu_1 & 0 & 0 & 0 & 0 & 0 \\ 0 & \gamma_1 & -\gamma_2 - \mu_2 & 0 & 0 & 0 & 0 \\ 0 & 0 & \gamma_2 & -\gamma_3 - \mu_3 & 0 & 0 & 0 \\ 0 & 0 & 0 & \gamma_3 & -\gamma_4 - \mu_4 & 0 & 0 \\ 0 & 0 & 0 & 0 & \gamma_4 & -\gamma_5 - \mu_5 & 0 \\ 0 & 0 & 0 & 0 & 0 & r\gamma_5 & -\mu_M \end{bmatrix},$$

$$E = \begin{bmatrix} -1 \\ 1 \\ 0 \\ 0 \\ 0 \\ 0 \\ -1 \end{bmatrix}, \quad e_1 = \begin{bmatrix} 1 \\ 0 \\ 0 \\ 0 \\ 0 \\ 0 \\ 0 \end{bmatrix}, \quad C = \begin{bmatrix} 0 & 1 & 1 & 0 & 0 & 0 & 0 \end{bmatrix}$$

Due to the dimension of the dynamics, it is not easy to check the identifiability of the parameter β. However, on the domain $X = (\mathbb{R}_+ \setminus \{0\})^7$, one has $SM \neq 0$, which implies the property

$$f_1(x, \beta_1) = f_1(x, \beta_2) \Rightarrow \beta_1 = \beta_2, \quad x \in X.$$

For the parameter $\theta = \beta$, condition (2.11) of Proposition 2.6 is thus fulfilled. Note that one has $ECE = E$. Therefore one can consider the variable

$$w = g(x, y) := x - Ey = (I - EC)x$$

whose dynamics is independent of the non-linear term βSM:

$$\dot{w} = l(w, y) := \bar{A}w + \bar{A}Ey + \Lambda e_1$$

where we posit $\bar{A} = A - ECA$. Given w and y, the state x is then given by

$$x = \tilde{g}(w, y) := w + Ey.$$

Conditions (2.12) and (2.13) of Proposition 2.6 are thus also satisfied, which allows to conclude without any other calculation that the parameter β is identifiable.

This example illustrates the possible interest of exploiting conjointly identifiability and observability to solve the identifiability problem.

Chapter 3
Analysis of the Kermack-McKendrick Model

Abstract This chapter is dedicated to the study of the well-known SIR model.

3.1 History

The SIR model of Kermack and McKendrick [73] is certainly one of the most famous models in Epidemiology. It is given and studied in all of the classic books of Mathematical Epidemiology. This model appears in the book of Bailey, which is probably the first book in Mathematical Epidemiology. Some examples can be found in [7, 24–26, 40, 80, 86, 89]. The figure, in the original paper, fitting the model to plague data in Bombay during the 1906 year, is one of the most famous pictures in Epidemiology. A research with SIR in `MathScinet` returns $11,106$ articles.

In the quoted books the SIR model is fitted to data in the following cases:

- in [24–26] the model is fitted to the plague in Eyam (in the year 1666);
- in [40] the model is fitted to an influenza epidemic in England and Wales;
- in [80] a fitting is done with simulated noisy data;
- in [24, 86], in a chapter devoted to fitting epidemiological models to data, a SIR model is fitted to an influenza outbreak in an English boarding school.

More recently two publications [12, 85] revisit the fit of the Kermack-McKendrick SIR model to the plague in Bombay.

As already mentioned, before attempting to adjust parameters, an identifiability analysis should be performed.

© The Author(s), under exclusive license to Springer Nature Singapore Pte Ltd. 2024
N. Cunniffe et al., *Identifiability and Observability in Epidemiological Models*,
SpringerBriefs on PDEs and Data Science,
https://doi.org/10.1007/978-981-97-2539-7_3

3.2   The Different Forms of the SIR Model

The original model [73] is

$$\begin{cases} \dot{S} = -\tilde{\beta}\,S\,I \\ \dot{I} = \tilde{\beta}\,S\,I - \gamma\,I \\ \dot{R} = \gamma\,I \end{cases} \tag{3.1}$$

where S, I, R represent respectively the numbers of susceptible, infectious and removed individuals.

This model can also be found in a slightly different form

$$\begin{cases} \dot{S} = -\beta\,\dfrac{S}{N}\,I \\ \dot{I} = \beta\,\dfrac{S}{N}\,I - \gamma\,I \\ \dot{R} = \gamma\,I \end{cases} \tag{3.2}$$

where $N = S + I + R$ is the total population. Obviously, one can pass from one model to the other though $\tilde{\beta} = \beta/N$. Both models are mathematically equivalent as long as N is a constant. However, we stress that identifying $\tilde{\beta}$ does not allow one to estimate the parameters β and N separately. For instance, estimating $\tilde{\beta}$ and γ only (without knowing N or β) does not allow one to estimate the basic reproduction number:

$$\mathcal{R}_0 = \frac{\tilde{\beta}\,N}{\gamma} = \frac{\beta}{\gamma}.$$

3.3   Observability and Identifiability of the SIR Model

Quite surprisingly, the observability and identifiability of the original Kermack-Mckendrick SIR model has not been studied much, although this model is commonly used to model epidemics.

Interestingly, the observability and identifiability of the SIR model with births and deaths, constant population, and an observation $y = k\,I$, has been first studied in 2005 [51]:

$$\begin{cases} \dot{S} = \mu\,N - \beta\,\dfrac{S}{N}\,I - \mu\,S \\ \dot{I} = \beta\,\dfrac{S}{N}\,I - (\gamma + \mu)\,I \\ \dot{R} = \gamma\,I - \mu\,R \end{cases} \tag{3.3}$$

where μ is the renewal rate of the population. The article [51] concludes that the system is neither observable nor identifiable.

In [120] the identifiability of (3.2) is addressed assuming (i) that the initial conditions (and therefore $N = S(0) + I(0) + R(0)$) are known, (ii) observing $y = kI$ with $k = 1$, and using only the input-output relation to conclude. Under assumptions (i) and (ii), the identifiability is quite immediate, as we shall see, but of limited interest.

3.3.1 The SIR Model When Observing a Ratio of the Infected Population

Here we study the observability-identifiability property of the SIR model

$$
\begin{cases}
\dot{S} = -\beta \dfrac{S}{N} I \\[2mm]
\dot{I} = \beta \dfrac{S}{N} I - \gamma I \\[2mm]
y = k I
\end{cases}
\tag{3.4}
$$

The dynamics of R has been omitted since $R = N - S - I$. The observation is $y = k I$ (with $k \in (0, 1)$), in other words only a fraction of the infectious individuals are observed. This situation is used for example in [85, 106]. In general, the values of N and k are also not known. We have the following result.

Theorem 3.1 *System (3.4) is neither observable, nor identifiable.*

Remark 3.1 Theorem 3.1 can be obtained from [51] by setting $\mu = 0$. However we provide a short and elementary proof.

Proof System (3.4) is obviously not observable on the invariant set $I = 0$. Therefore we study the observability-identifiability properties of System (3.4) on the following positively invariant open set

$$
\mathcal{D} = \{(S, I) \mid S > 0, \ I > 0, \ S + I < N\}.
$$

We will show that there exist a couple of distinct initial condition and distinct parameters that generate the same output. The computation of successive time derivatives of the output y gives:

$$\begin{cases} y = kI, \\[2mm] \dot{y} = y\left(\dfrac{\beta S}{N} - \gamma\right), \\[3mm] \ddot{y} = -\dfrac{\beta^2 S}{kN^2}y^2 + \dfrac{\dot{y}^2}{y}, \\[3mm] \dddot{y} = \left(-y\ddot{y} + \dot{y}^2\right)\dfrac{\beta}{kN} - \dfrac{3\dot{y}^3}{y^2} + \dfrac{4\dot{y}\ddot{y}}{y}. \end{cases} \qquad (3.5)$$

One can observe that the third derivative of y is expressed as a function of lower derivatives of y of and parameter $\frac{\beta}{kN}$ only. This property remains true for any further derivative of y. This implies that if one considers two sets of initial condition and parameters such that the expressions of y, $\dot{y}$, $\ddot{y}$ and $\frac{\beta}{kN}$ coincide, then any further derivative also coincides. By analyticity of the solutions of the system as well as of the corresponding outputs, we deduce that their outputs are the same for any time $t > 0$. More precisely, take two different initial conditions $(S_0, I_0) \neq (\bar{S}_0, \bar{I}_0)$ and sets of parameters $(N, k, \gamma, \beta) \neq (\bar{N}, \bar{k}, \bar{\gamma}, \bar{\beta})$ such that

$$kI_0 = \bar{k}\bar{I}_0,$$

$$\frac{\beta}{kN} = \frac{\bar{\beta}}{\bar{k}\bar{N}},$$

$$\frac{\beta S_0}{N} - \gamma = \frac{\bar{\beta}\bar{S}_0}{\bar{N}} - \bar{\gamma},$$

$$\frac{\beta^2 S_0}{kN^2} = \frac{\bar{\beta}^2 \bar{S}_0}{\bar{k}\bar{N}^2}$$

then $x_0 = (S_0, I_0, N, k, \gamma, \beta)$ and $\bar{x}_0 = (\bar{S}_0, \bar{I}_0, \bar{N}, \bar{k}, \bar{\gamma}, \bar{\beta})$ generate the same output function $y(\cdot)$.   □

Therefore, system (3.4) is not observable nor identifiable. However, it is *partially observable* (respectively *partially* identifiable) in the sense of Definitions 1.4 and 1.5 that is some functions of the state variables (respectively of the parameters) are observable (respectively identifiable). This is made precise in the following proposition.

Proposition 3.1 *System (3.4) satisfies the following properties:*

i. *The state functions* kI, $\dfrac{\beta S}{N}$ *and* kS *are observable.*

ii. *The parameter* γ *and the parameter function* $\dfrac{\beta}{k\,N}$ *are identifiable.*

Proof From (3.5), and since one has $-y\ddot{y} + \dot{y}^2 = \dfrac{\beta^2 k^2 I^3 S}{N^2} > 0$, we get

$$\begin{cases} kI = y, \\[2mm] \dfrac{\beta S}{N} - \gamma = \dfrac{\dot{y}}{y}, \\[2mm] \dfrac{\beta^2 S}{k N^2} = \dfrac{\dot{y}^2}{y^3} - \dfrac{\ddot{y}}{y^2}, \\[2mm] \dfrac{\beta}{kN} = \dfrac{\dddot{y} - \dfrac{4\dot{y}\ddot{y}}{y} + \dfrac{3\dot{y}^3}{y^2}}{-y\ddot{y} + \dot{y}^2}. \end{cases}$$

From these expressions, we deduce

$$\frac{\beta S}{N} = \frac{(y\ddot{y} - \dot{y}^2)^2}{y\left(\dddot{y}\, y^2 - 4\dot{y}\ddot{y}y + 3\dot{y}^3\right)},$$

$$\gamma = \frac{(y\ddot{y} - \dot{y}^2)^2}{y\left(\dddot{y}\, y^2 - 4\dot{y}\ddot{y}y + 3\dot{y}^3\right)} + \frac{\dot{y}}{y},$$

$$kS = \frac{\dfrac{\beta S}{N}}{\dfrac{\beta}{kN}} = \frac{\left(-y\ddot{y} + \dot{y}^2\right)^3 y}{\left(y^2\dddot{y} - 4\dot{y}\ddot{y}y + 3\dot{y}^3\right)^2}.$$

Thus the parameters γ and $\dfrac{\beta}{k N}$ are identifiable and the state functions kI, $\dfrac{\beta S}{N}$ and kS are observable.   □

Remark 3.2 Since the total population N (constant for the model considered) is often considered known in epidemiology or epidemic modeling (e.g. from the Census Bureau), Proposition 3.1 implies that the quantity β/k is identifiable. However β and k are not identifiable independently which means that there are infinitely many combinations of β and k values for which the model produce the same observable output.

As a consequence of Proposition 3.1, we have also the following properties.

Corollary 3.1 *For system* (3.4), *one has*

i. *If $k = \gamma$, then the state variables S and I are observable, and the parameters γ, $\dfrac{\beta}{N}$ are identifiable.*

ii. *If N is known and if $k = 1$ or $k = \gamma$ then the system is identifiable and observable.*

Remark 3.3 One could believe at the first look that if $k = \gamma$ but with N unknown, then (3.1) is observable, but this is wrong. Certainly S and I are observable, but R is not observable. Indeed the output $y = kI$ and state I, solution of system (3.1), do not depend on the variable R since R has no influence on the two first equations of (3.1). So, the output generated by system (3.1) will be the same for any initial conditions (S_0, I_0, R_0) and $(S_0, I_0, \bar{R}_0)$ even if $R_0 \neq \bar{R}_0$. Therefore the value of $N = S + I + R$ is inaccessible. As a consequence, $\mathcal{R}_0 = \tilde{\beta} N / \gamma$ is not identifiable.

Remark 3.4 The second point of Corollary 3.1 can be also proved by considering the "augmented" system (defined in Chap. 1 by Eq. (1.3)) and using Proposition 1.1. Indeed, when N is known and $k = \gamma$, consider the "augmented" SIR model as follows

$$\begin{cases} \dot{S} = -\beta \dfrac{S}{N} I \\ \dot{I} = \beta \dfrac{S}{N} I - \gamma I \\ \dot{\beta} = 0 \\ \dot{\gamma} = 0 \\ \\ y = \gamma I \end{cases} \tag{3.6}$$

Posit $x = \begin{bmatrix} S \\ I \\ \beta \\ \gamma \end{bmatrix}$, $f(x) = \begin{bmatrix} -\beta \dfrac{S}{N} I \\ \beta \dfrac{S}{N} I - \gamma I \\ 0 \\ 0 \end{bmatrix}$ and $h(x) = \gamma I$. We have

$$\mathcal{L}_f h(x) = \langle \nabla h(x) | f(x) \rangle = \langle \begin{bmatrix} 0 \\ \gamma \\ 0 \\ I \end{bmatrix} \Big| \begin{bmatrix} -\beta \dfrac{S}{N} I \\ \beta \dfrac{S}{N} I - \gamma I \\ 0 \\ 0 \end{bmatrix} \rangle$$

$$= \gamma \left(\beta \dfrac{S}{N} I - \gamma I \right) = \gamma \left(\beta \dfrac{S}{N} - \gamma \right) I = \left(\beta \dfrac{S}{N} - \gamma \right) h(x).$$

$$\mathcal{L}_f^2 h(x) = \langle \nabla \mathcal{L}_f h(x) | f(x) \rangle = \langle \begin{bmatrix} \gamma \dfrac{\beta}{N} I \\ \gamma \left(\beta \dfrac{S}{N} - \gamma \right) \\ \gamma \dfrac{S}{N} I \\ \beta \dfrac{S}{N} I \end{bmatrix} \Big| \begin{bmatrix} -\beta \dfrac{S}{N} I \\ \beta \dfrac{S}{N} I - \gamma I \\ 0 \\ 0 \end{bmatrix} \rangle$$

$$= -\gamma \frac{\beta^2}{N^2} SI^2 + \gamma \left(\beta \frac{S}{N} - \gamma \right)^2 I$$

$$= -\gamma \frac{\beta^2}{N^2} SI^2 + \left(\beta \frac{S}{N} - \gamma \right) \mathcal{L}_f h(x)$$

$$= -\frac{\beta^2}{N^2} SI\, h(x) + \left(\beta \frac{S}{N} - \gamma \right) \mathcal{L}_f h(x).$$

and computing the third Lie-derivative of h gives the expression

$$\mathcal{L}_f^3 h(x)$$

$$= -\frac{\beta^2}{N^2} SI \left(\frac{\beta S}{N} - \gamma - \frac{\beta I}{N} \right) h(x) - 2\frac{\beta^2}{N^2} SI\, \mathcal{L}_f h(x) + \left(\frac{\beta S}{N} - \gamma \right) \mathcal{L}_f^2 h(x)$$

$$\mathcal{L}_f^3 h(x)$$

$$= \frac{\left(S \left(I^2 - 4IS + S^2 \right) \beta^3 + (4I - 3S)\, S\gamma N\, \beta^2 + 3N^2 S\beta\, \gamma^2 - N^3\gamma^3 \right) \gamma\, I}{N^3}.$$

We show that the map

$$x \mapsto H_3(x) = (h(x), \mathcal{L}_f h(x), \mathcal{L}_f^2 h(x), \mathcal{L}_f^3 h(x))^\top \tag{3.7}$$

is injective. Let x and $\bar{x}$ two elements of the state space $\Omega = \{(S, I, \beta, \gamma) \in \mathbb{R}^4 : S > 0, I > 0, \beta > 0, \gamma > 0\}$. Suppose that $H_3(x) = H_3(\bar{x})$. Then, we have the following successive implications:
$$h(x) = h(\bar{x}), \mathcal{L}_f(h)(x) = \mathcal{L}_f(h)(\bar{x}), \mathcal{L}_f^2(h)(x) = \mathcal{L}_f^2(h)(\bar{x}) \implies$$

$$\begin{cases} \gamma I = \bar{\gamma} \bar{I} \\ \left(\beta \frac{S}{N} - \gamma \right) h(x) = \left(\bar{\beta} \frac{\bar{S}}{N} - \bar{\gamma} \right) h(\bar{x}) \\ -\frac{\beta^2}{N^2} SI\, h(x) + \left(\beta \frac{S}{N} - \gamma \right) \mathcal{L}_f h(x) = -\frac{\bar{\beta}^2}{N^2} \bar{S}\bar{I}\, h(\bar{x}) + \left(\bar{\beta} \frac{\bar{S}}{N} - \bar{\gamma} \right) \mathcal{L}_f h(\bar{x}) \end{cases}$$

implies

$$
\begin{cases}
\gamma I = \bar{\gamma}\bar{I} \\
\left(\beta\dfrac{S}{N} - \gamma\right) = \left(\bar{\beta}\dfrac{\bar{S}}{N} - \bar{\gamma}\right) \\
\dfrac{\beta^2}{N^2}SI = \dfrac{\bar{\beta}^2}{N^2}\bar{S}\bar{I}
\end{cases}
\implies
\begin{cases}
\gamma I = \bar{\gamma}\bar{I} \\
\left(\beta\dfrac{S}{N} - \gamma\right) = \left(\bar{\beta}\dfrac{\bar{S}}{N} - \bar{\gamma}\right) \\
\beta^2 SI = \bar{\beta}^2\bar{S}\bar{I}
\end{cases}
$$

$$
\implies
\begin{cases}
\gamma I = \bar{\gamma}\bar{I} \\
\beta\dfrac{S}{N} - \gamma = \bar{\beta}\dfrac{\bar{S}}{N} - \bar{\gamma} \\
\beta^2 S = \bar{\beta}^2\dfrac{\gamma}{\bar{\gamma}}\bar{S}
\end{cases}
\implies
\begin{cases}
\gamma I = \bar{\gamma}\bar{I} \\
\beta\dfrac{S}{N} - \gamma = \dfrac{\beta^2\bar{\gamma}}{\bar{\beta}\gamma}\dfrac{S}{N} - \bar{\gamma} \\
\beta^2 S = \bar{\beta}^2\dfrac{\gamma}{\bar{\gamma}}\bar{S}
\end{cases}
$$

Reporting in $\mathcal{L}_f^3(h)(x) = \mathcal{L}_f^3(h)(\bar{x})$, we obtain

$$
-\frac{\beta^2}{N^2}SI\left(\frac{\beta S}{N} - \gamma - \frac{\beta I}{N}\right)h(x) - 2\frac{\beta^2}{N^2}SI\,L_f h(x) + \left(\frac{\beta S}{N} - \gamma\right)L_f^2 h(x)
$$

$$
= -\frac{\beta^2}{N^2}SI\left(\frac{\beta S}{N} - \gamma - \frac{\bar{\beta}\bar{I}}{N}\right)h(x) - 2\frac{\beta^2}{N^2}SI\,L_f h(x) + \left(\frac{\beta S}{N} - \gamma\right)L_f^2 h(x)
$$

This implies

$$
\frac{\beta I}{N} = \frac{\bar{\beta}\bar{I}}{N} \text{ and } \beta I = \bar{\beta}\bar{I}
$$

With $\gamma I = \bar{\gamma}\bar{I}$, we obtain $\bar{\beta} = \dfrac{\gamma}{\bar{\gamma}}\beta$. Reporting in $\beta\dfrac{S}{N} - \gamma = \dfrac{\beta^2\bar{\gamma}}{\bar{\beta}\gamma}\dfrac{S}{N} - \bar{\gamma}$, we obtain $\bar{\gamma} = \gamma$. We then deduce $\bar{I} = I$, $\bar{\beta} = \beta$, and $\bar{S} = S$, and thus $x = \bar{x}$. We have shown that the map H_3 (3.7) is injective which proves that the augmented system (3.6) is observable which implies, thanks to Proposition 1.1, that System (3.4) is observable and identifiable when the total (constant) population N is known and $k = \gamma$.

3.3.2   The SIR Model When Observing the Incidence

Quite often, observations of new cases per unit time or incidence are available. We study how this changes the observability and identifiability of the SIR model.

We thus consider the system where the observation is given by

$$y = k\beta \frac{SI}{N}. \tag{3.8}$$

The system under consideration is then

$$\begin{cases} \dot{S} = -\beta \dfrac{S}{N} I, \\[2mm] \dot{I} = \beta \dfrac{S}{N} I - \gamma I, \\[2mm] y = k\beta \dfrac{SI}{N}. \end{cases} \tag{3.9}$$

This problem has been addressed for the SIR model with demography for constant population in [51]. Identifiability with known initial conditions for (3.9) is also considered in [120] using input-output relations.

Theorem 3.2 *The system (3.9) with incidence observation is neither observable, nor identifiable.*

Proof We proceed as we did for the proof of Theorem 3.1. The computation of successive time derivatives of the output y gives:

$$\begin{cases} y & = k\beta \dfrac{SI}{N}, \\[3mm] \dot{y} & = \left(\dfrac{\beta S}{N} - \dfrac{\beta I}{N} - \gamma \right) y, \\[3mm] \ddot{y} & = \dfrac{\dot{y}^2}{y} + \left(-\dfrac{2\beta}{kN} y + \dfrac{\gamma \beta I}{N} \right) y, \\[3mm] y^{(3)} & = \dfrac{2\dot{y}\ddot{y}}{y} - \dfrac{\dot{y}^3}{y^2} - 4\dfrac{\beta}{kN} y\dot{y} + \dfrac{\gamma\beta I}{N}\dot{y} + \dfrac{\gamma\beta}{N} y \left(\dfrac{y}{k} - \gamma I \right) \\[3mm] & = \dfrac{2\dot{y}\ddot{y}}{y} - \dfrac{\dot{y}^3}{y^2} + \dfrac{\beta}{kN} \left(-4y\dot{y} + \gamma y^2 \right) + \dfrac{\gamma\beta I}{N} (\dot{y} - \gamma y) \end{cases} \tag{3.10}$$

From the expression of $\ddot{y}$, one takes

$$\frac{\gamma\beta I}{N} = \frac{\ddot{y}}{y} - \frac{\dot{y}^2}{y^2} + 2\frac{\beta}{kN} y$$

and reporting in the expression of $y^{(3)}$ gives

$$
\begin{aligned}
y^{(3)} &= \frac{2\dot{y}\ddot{y}}{y} - \frac{\dot{y}^3}{y^2} + \frac{\beta}{kN}\left(-4y\dot{y} + \gamma y^2\right) + \left(\frac{\ddot{y}}{y} - \frac{\dot{y}^2}{y^2} + 2\frac{\beta}{kN}y\right)(\dot{y} - \gamma y) \\
&= \frac{2\dot{y}\ddot{y}}{y} - \frac{\dot{y}^3}{y^2} + \left(\frac{\ddot{y}}{y} - \frac{\dot{y}^2}{y^2}\right)(\dot{y} - \gamma y) + \frac{\beta}{kN}\left(-4y\dot{y} + \gamma y^2 + 2y(\dot{y} - \gamma y)\right) \\
&= \frac{2\dot{y}\ddot{y}}{y} - \frac{\dot{y}^3}{y^2} + \left(\frac{\ddot{y}}{y} - \frac{\dot{y}^2}{y^2}\right)(\dot{y} - \gamma y) + \frac{\beta}{kN}\left(-2y\dot{y} - \gamma y^2\right) \\
&= \frac{\ddot{y}\dot{y}}{y} + \left(\frac{2\dot{y}}{y} - \gamma\right)\frac{\left(y\ddot{y} - \dot{y}^2\right)}{y} - \frac{\beta}{kN}\left(\gamma y + 2\dot{y}\right)y.
\end{aligned}
$$

To summarize, the successive derivatives are

$$
\begin{cases}
y = k\beta\,\dfrac{S\,I}{N}, \\[2mm]
\dot{y} = \left(\dfrac{\beta S}{N} - \dfrac{\beta I}{N} - \gamma\right)y, \\[2mm]
\ddot{y} = \dfrac{\dot{y}^2}{y} + \left(-\dfrac{2\beta}{kN}y + \dfrac{\gamma\beta I}{N}\right)y, \\[2mm]
y^{(3)} = \dfrac{\ddot{y}\dot{y}}{y} + \left(\dfrac{2\dot{y}}{y} - \gamma\right)\dfrac{\left(y\ddot{y} - \dot{y}^2\right)}{y} - \dfrac{\beta}{kN}\left(\gamma y + 2\dot{y}\right)y, \\[2mm]
y^{(3)} = \dfrac{3y\dot{y}\ddot{y} - 2\dot{y}^3}{y^2} + \dfrac{\dot{y}^2 - y\ddot{y}}{y}\gamma - \dfrac{\beta}{kN}\left(\gamma y + 2\dot{y}\right)y
\end{cases}
\tag{3.11}
$$

This shows that all the derivatives of y of order $q \geq 3$ can be expressed as functions of derivatives $y^{(r)}$, $r = 0, \cdots, q - 1$ and parameters γ, $\dfrac{\beta}{kN}$, only. We deduce that for two different $x = (S_0, I_0, N, k, \gamma, \beta)$ and $\bar{x} = (\bar{S}_0, \bar{I}_0, \bar{N}, \bar{k}, \bar{\gamma}, \bar{\beta})$ such that

$$
\gamma = \bar{\gamma} \quad \text{and} \quad \frac{\beta}{kN} = \frac{\bar{\beta}}{\bar{k}\bar{N}}
$$

and derivatives $y^{(r)}(0)$, $r = 0 \cdots 2$, coincide, then their output $y(t)$ is the same for any $t > 0$ (by analyticity of the system). Notice that one can write

$$
y = k\beta\frac{S\,I}{N} = \frac{\beta I}{N}\frac{\beta S}{N}\left(\frac{\beta}{kN}\right)^{-1}.
$$

Therefore, from expressions (3.11), $y^{(r)}(0)$, $r \in \{0, 1, 2\}$ coincide under the single condition

$$\frac{\beta I_0}{N} = \frac{\bar{\beta} \bar{I}_0}{\bar{N}}, \quad \text{and} \quad \frac{\beta S_0}{N} = \frac{\bar{\beta} \bar{S}_0}{\bar{N}}.$$

We conclude that the system is neither observable, nor identifiable.   □

Proposition 3.2 *If the parameters N, γ and k are known than System (3.9) is observable and identifiable.*

Proof Thanks to Proposition 1.1, to prove that System (3.9) is observable and identifiable, it is sufficient to prove that the following (augmented) system whose state is (S, I, β):

$$\begin{cases} \dot{S} = -\beta \dfrac{S}{N} I \\[2mm] \dot{I} = \beta \dfrac{S}{N} I - \gamma I \\[2mm] \dot{\beta} = 0 \\[2mm] y = k\beta \dfrac{SI}{N} \end{cases} \tag{3.12}$$

is observable. From the expression of $y^{(3)}$, we obtain

$$\beta = \frac{\gamma(\dot{y}^2 - y\ddot{y})y + (3\dot{y}\ddot{y} - yy^{(3)})y - 2\dot{y}^3}{\gamma y^4 + 2\dot{y}y^3} kN.$$

Reporting in $\ddot{y}$, we obtain I as a function of N, γ, k, and the successive derivatives of y. The state variable S is obtained by reporting the expressions of β and I in the expression of y. Thus System (3.12) satisfies the condition of Proposition 2.3 and therefore it is observable on $\{(S, I, \beta) \in \mathbb{R}^3 : S > 0, I > 0, S + I < N, \beta > 0\}$.

□

Remark 3.5 If it is rather the cumulative incidence that is observed, that is

$$y(t) = k \int_0^t \beta \frac{S(\tau) I(\tau)}{N} d\tau.$$

the system is also not observable or identifiable, and the same results as in Proposition 3.2 are available. Indeed all the k-th derivatives of y (for $k \geq 1$) correspond to the $k - 1$-th derivatives of the observation (3.8).

Chapter 4
Observers Synthesis

Abstract We expose the concept of observers and review different techniques to build observers.

4.1 Introduction

As in the previous chapters we consider an observed system

$$\begin{cases} \dot{x} = f(x), & x \in \Omega \subset \mathbb{R}^n, \\ y = h(x) \in Y \subset \mathbb{R}^m \end{cases} \tag{4.1}$$

and denote by $x(t, x_0)$ the solution of $\dot{x} = f(x)$ for the initial condition $x(0) = x_0$.

So far we have studied observability as a property ensuring that knowledge of a measured "signal" $y(\cdot)$ results in the uniqueness of the initial condition x_0. In this chapter we will address the *state estimation problem* which consists in obtaining an estimate $\hat{x}(t)$ of the state $x(t)$ of the system at time t, with the knowledge of the output $y(\cdot)$ up to time t.

We have for addressing this problem different possibilities.

The first one is *given the output $y(\cdot)$ up to time t, find the possible initial condition x_0 which produces the same output $y(\cdot)$ up to time t*. Then, the estimate of $x(t)$ is given by the solution $x(t, x_0)$. The uniqueness of this problem is ascertained by the observability property of the system. This approach leads to the resolution of a *minimization problem*.

$$\min_{x_0} \int_0^t \| h(x(s, x_0)) - y(s) \|^2 \, ds.$$

In other words, we look, by simulating the system for different initial conditions, for the "best" one. The drawback of this approach is related to the difficulties of the nonlinear minimization algorithms (existence of different local minimal, convergence

© The Author(s), under exclusive license to Springer Nature Singapore Pte Ltd. 2024
N. Cunniffe et al., *Identifiability and Observability in Epidemiological Models*,
SpringerBriefs on PDEs and Data Science,
https://doi.org/10.1007/978-981-97-2539-7_4

speed...). Once one obtains an estimation $\hat{x}_0$ of the solution of this minimization problem, the estimation of the state $x(t)$ is given by $\hat{x}(t) = x(t, \hat{x}_0)$.

A second approach is to differentiate the available outputs a number of times and then combine these derivatives appropriately to obtain the state vector. Formally, when the system is observable and we know (perfectly) enough derivatives of $y(\cdot)$ at a given time t, one just has to invert the map $\phi \mapsto (h(x), \mathcal{L}_f h(x), \cdots)$ at $(y(t), \dot{y}(t), \ddot{y}(t), \cdots)$ to reconstruct the state variable $x(t)$. In practice, it is known that numerically calculating derivatives on the raw signal data is imprecise and sensitive to measurement noise, especially if several successive derivatives have to be determined. It is generally preferable to use a "filter" to smooth the data. For instance, polynomial functions or splines can approach with some regularity the measurements obtained over time, on which the derivative calculations can be performed before the inversion operation. This is an approximation method that does not guarantee an exact solution, and whose accuracy can be strongly influenced by the sensitivity of the solutions of the system with respect to the derivatives of the output (and the smoothing).

The last approach is to look for a dynamical system whose "inputs" are the output of the observed system y, and whose "output" is an estimate $\hat{x}$ of the state of the original system, illustrated on the diagram of Fig. 4.1. Such dynamical system is called an observer and is classical in the theory of control.

Definition 4.1 An observer for system (4.1) is an input-output system of the form

$$\begin{cases} \dot{\hat{z}} = g(\hat{z}, y(t)), \ \hat{z} \in Z \subset \mathbb{R}^{n_z} \\ \hat{x} = l(\hat{z}, y(t)) \end{cases} \tag{4.2}$$

such that $\hat{x}(t)$ is an asymptotic estimate of $x(t)$ satisfying

$$\lim_{t \to +\infty} ||\hat{x}(t) - x(t)|| = 0$$

for any initial condition $(x(0), \hat{z}(0)) \in \Omega \times Z$.

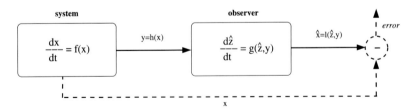

Fig. 4.1 The observer concept

In this definition, note that n_z is not necessarily equal to n. It can be less than n for *reduced-order* observers (see Sect. 4.5 below) or larger than n (see Remark 4.3 later on). Let us also underline that this amounts to require the convergence of the solution of the coupled dynamics

$$\begin{cases} \dot{x} = f(x) \\ \dot{\hat{z}} = g(\hat{z}, h(x)) \end{cases} \Rightarrow \lim_{t \to +\infty} l(\hat{z}(t), h(x)) - x(t) = 0$$

for any initial condition in $\Omega \times Z$.

In this Chapter, we study the construction of observers and their theoretical convergence, without considering their practical performances in presence of measurements noise. This point will be addressed with more practical considerations in Chap. 5.

Many observers are indeed of the form

$$\dot{\hat{x}} = f(\hat{x}) + G(h(\hat{x}) - y(t))$$

where G is a constant matrix (i.e. of the form (4.2) with $\hat{z} = \hat{x}$). Such observers are often called *Luenberger observers* [84]. Note that this construction consists in a copy of the original dynamics f plus a correcting term which depends on the *innovation term* $h(\hat{x}) - y(t)$, *that is defined as the difference between the expected output* $\hat{y} = h(\hat{x}(t)))$ *if the true state was* $\hat{x}(t)$ *and the effective measured output* $y(t)$. Therefore, if $\hat{x}(t)$ and $x(t)$ are equal at a certain time t, it will remain identical at any future time. The main point is that the matrix G, often called the *gains matrix*, has to be chosen such that the estimation error $\hat{x}(t) - x(t)$ does converge to 0, possibly fast. When f and h are linear and the system is observable, the theory of linear automatic control teaches that there always exists G such that the convergence speed of the estimator can be chosen arbitrarily fast (see e.g. [8]). Obviously epidemiological models are rarely linear. However, looking for a Luenberger observer is often a first trial before considering more sophisticated estimators. Indeed, we shall see that for certain nonlinear dynamics, such observers do the job and in other cases, observers can be inspired from this form.

In this chapter, we do not pretend to present an exhaustive review of all the possible kinds of observers that exist in the literature (we make some comments at the end of the chapter). We focus on the most general classes of observers, for which the proof of the convergence can be shown, under some assumptions, and that can be derived in a systematic way from the equations of the model, once theses assumptions are fulfilled. We begin by some simple cases of observers for particular dynamics exploiting properties of linear systems, and then we consider a more general non-linear framework. The numerical implementation of these observers is illustrated in Chap. 5.2.

4.2   Observers with Linear Error Dynamics

Consider systems of the form

$$
\begin{cases}
\dot{x} = Ax + \Phi(t, y) \\
y = Cx
\end{cases}
\tag{4.3}
$$

where $x \in \Omega$ and $y \in \mathbb{R}$. We are looking for observers in the Luenberger form

$$
\dot{\hat{x}} = A\hat{x} + \Phi(t, y(t)) + G(C\hat{x} - y(t)), \quad \hat{x} \in \mathbb{R}^n
\tag{4.4}
$$

for which the error vector $e(t) = \hat{x}(t) - x(t)$ is solution of the linear dynamics

$$
\dot{e} = (A + GC)e
\tag{4.5}
$$

The choice of the *gains* vector G providing a convergence of $e(t)$ to 0 comes directly from *the poles placement* technique of the theory of linear systems that we recall below. Let us underline that although Ω is assumed to be positively invariant by the dynamics (4.3), there is no reason for Ω to be invariant by (4.4) because of the additive correction term. This is why we consider the dynamics (4.4) in whole $\mathbb{R}^n$.

   The following lemma gives a key result to construct the matrix G, which is well-known and often used in automatic control.

Lemma 4.1 *Let A be a square matrix of size n and C be a line vector of length n. If the observability matrix O defined in (2.4) is of full rank, then for any set $\Lambda = \{\lambda_1, \cdots, \lambda_n\}$ of n real or complex two-by-two conjugate numbers, there is a G vector of size n such that*

$$
Sp(A + GC) = \Lambda .
$$

Specifically, if

$$
\pi_A(\xi) = \xi^n + a_1\xi^{n-1} + \cdots + a_{n-1}\xi + a_n
$$

is the characteristic polynomial of A, then one has

$$
G = P\Big[a_{n+1-i} + (-1)^{n-i}\sigma_{n+1-i}(\Lambda)\Big]_{i=1}^n
$$

where

$$P = O^{-1} \begin{bmatrix} 0 \\ \vdots \\ \vdots \\ 1 \end{bmatrix} [I \ A \ \cdots \ A^{n-1}]$$

and the σ_k designate the symmetric functions of the roots

$$\sigma_k(\Lambda) = \sum_{1 \le i_1 \le \cdots \le i_k \le n} \lambda_{i_1} \lambda_{i_2} \cdots \lambda_{i_k} \qquad (4.6)$$

A proof of Lemma 4.1 is given in Appendix A.1.

Remark This result can be generalized to vectorial observations, i.e. for matrices C with $m > 1$ rows and n columns.

Finally, by choosing numbers λ_i with negative real parts, one can make the convergence of the error given by the exponentially decreasing dynamics (4.5) as fast as desired. The observer (4.4) is thus adjustable. It should be noted that when the difference $C\hat{x}(t) - y(t)$ (the "innovation") becomes and remains close to 0, the trajectories of the observer follow those of the system: we can then consider that the observer has practically converged. The innovation is thus very useful in practice because it provides information on the current stage of convergence of the estimate.

We illustrate this technique on a population model with age classes.

Example 4.1 Let us consider a population structured in three stages: young, subadult and adult, of stocks x_1, x_2, x_3 respectively. It is assumed that only adults x_3 can reproduce, giving birth to young x_1:

$$\begin{cases} \dot{x}_1 = -a_1 x_1 - m_1 x_1 + r(t, x_3) \\ \dot{x}_2 = a_1 x_1 - a_2 x_2 - m_2 x_2 \\ \dot{x}_3 = a_2 x_2 - m_3 x_3 \\ \\ y = x_3 \end{cases} \qquad (4.7)$$

The coefficients a_i are the transition rates between age classes, m_i are the mortality rates of each class, and $r(\cdot)$ is the reproductive function (usually non-linear and seasonally dependent), for example

$$r(t, x_3) = \frac{\bar{r}(t) x_3}{k + x_3}, \quad \bar{r}(t) \in [\bar{r}_{min}, \bar{r}_{max}].$$

Here, it is assumed that only the size of the adult class is measured over time. The aim is to estimate the stocks of young and subadults over time. The model (4.7) is of the form (4.3) with $\Omega = \mathbb{R}_+^3$, where we posed

$$A = \begin{bmatrix} -(a_1 + m_1) & 0 & 0 \\ a_1 & -(a_2 + m_2) & 0 \\ 0 & a_2 & -m_3 \end{bmatrix}, \quad C = [0\ 0\ 1]$$

and

$$\Phi(t, x_3) = \begin{bmatrix} r(t, x_3) \\ 0 \\ 0 \end{bmatrix}$$

One can check that the observability matrix O si full rank, or alternatively directly check that the system is observable. Indeed, we obtain by using the expression $\dot{x}_3$:

$$x_2 = \frac{\dot{y} + m_3 y}{a_2},$$

then with the expression $\dot{x}_2$:

$$x_1 = \frac{\dot{x}_2 + (a_2 + m_2)x_2}{m_1} = \frac{\ddot{y} + (a_2 + m_2)m_3 y}{a_2 m_2}.$$

Therefore, the following system in $\mathbb{R}^3$

$$\begin{cases} \dot{\hat{x}}_1 = -a_1\hat{x}_1 - m_1\hat{x}_1 + r(t, y(t)) + G_1(\hat{x}_3 - y(t)) \\ \dot{\hat{x}}_2 = a_1\hat{x}_1 - a_2\hat{x}_2 - m_2\hat{x}_2 + G_2(\hat{x}_3 - y(t)) \\ \dot{\hat{x}}_3 = a_2\hat{x}_2 - m_3\hat{x}_3 + G_3(\hat{x}_3 - y(t)) \end{cases} \quad (4.8)$$

with well chosen gains G_1, G_2, G_3 is an observer for the dynamics (4.7), with an exponential convergence.

In the next section, we study a more general kind of non-linearity.

4.3   Observers for Systems with Lipschitz Non-linearity

Let us consider a system of the form

$$\begin{cases} \dot{x} = Ax + \phi(x) \\ y = Cx \end{cases} \quad (4.9)$$

where

1. the pair (A, C) is observable i.e. the observability matrix O defined in (2.4) is of full rank,
2. the map ϕ is globally Lipschitz on $\mathbb{R}^n$, with a Lipschitz constant ℓ.

For this non-linear system, consider an observer in Luenberger form

$$\dot{\hat{x}} = A\hat{x} + \phi(\hat{x}) + G(C\hat{x} - y(t)) \tag{4.10}$$

We recall the Lyapunov Theorem [97], which is useful in many situations.

Theorem 4.1 *Given a symmetric positive definite matrix Q, there exists a unique symmetric definite positive matrix P satisfying $M^\top P + PM + Q = 0$ if and only if the linear system $\dot{x} = Mx$ is globally exponentially stable.*

A proof of this Theorem is given in Appendix A.3. Then, one has the following result.

Proposition 4.1 *Take G such that $A + GC$ is Hurwitz,[1] then if ℓ is small enough, (4.10) is an exponential observer of (4.9), in the sense that there exist $\alpha > 0$ and $\beta > 0$ such that*

$$||\hat{x}(t) - x(t)|| \leq \alpha ||\hat{x}(0) - x(0)||e^{-\beta t}, \quad t \geq 0$$

for any initial condition $x(0)$, $\hat{x}(0)$.

Proof Accordingly to Theorem 4.1, there exists a unique symmetric definite matrix P satisfying

$$(A + GC)^\top P + P(A + GC) + I = 0 \tag{4.11}$$

(where I is the identity matrix). Let us consider the candidate Lyapunov function

$$V(e) = e^\top Pe$$

and the time function $v(t) = V(e(t))$. One gets

$$\dot{v} = e^\top \left((A + GC)^\top P + P(A + GC)\right)e + 2e^\top P(\phi(\hat{x}) - \phi(x))$$

$$= -||e||_2^2 + 2e^\top P(\phi(\hat{x}) - \phi(x))$$

$$\leq -||e||_2^2 + 2\ell||P||_2||e||_2^2 = -(1 - 2\ell||P||_2)||e||_2^2$$

---

[1] A matrix is Hurwitz if all its eigenvalues have negative real parts.

If the Lipschitz constant ℓ is small, one has

$$1 - 2\ell||P||_2 > 0. \tag{4.12}$$

On the other hand, as P is a positive definite matrix, one has

$$\lambda_{min}||e||_2 \leq V(e) \leq \lambda_{max}||e||_2$$

where λ_{min}, λ_{max} are the smallest and largest eigenvalues of P. Then one obtains $\dot{v} \leq -2\beta v$ where

$$\beta = \frac{1 - 2\ell||Pe||_2}{2\lambda_{min}} > 0$$

which implies

$$v(t) \leq v(0)e^{-2\beta t} \Rightarrow ||e(t)||^2 \leq \alpha^2||e(0)||^2 e^{-2\beta t}$$

with $\alpha = \sqrt{\frac{\lambda_{max}}{\lambda_{min}}}$. Finally, one gets

$$||e(t)|| \leq \alpha||e(0)||e^{-\beta t}, \quad t \geq 0$$

which proves that the estimation error e converges exponentially to 0.

There exist other techniques to design a gain G based on the Riccati equation rather than the Lyapunov equation (4.11), but that are more technical and thus out of the scope of this book (we refer to [2] for interested readers).

Remark 4.1 If the map ϕ can be written as $\phi(x) = \varphi(h(x), x) = \varphi(y, x)$, where φ is globally Lipschitz w.r.t. x uniformly in y, then one can consider the observer

$$\dot{\hat{x}} = A\hat{x} + \varphi(y(t), \hat{x}) + G(C\hat{x} - y(t)) \tag{4.13}$$

which generalizes the observers (4.4) and (4.10).

Example 4.2 We consider the "SIRS" model [74], that is the SIR model with loss of immunity, assuming that the size of the recovered population is tracked over time

$$\begin{cases} \dot{S} = -\beta SI + \mu R \\ \dot{I} = \beta SI - \gamma I \\ \dot{R} = \gamma I - \mu R \\ \\ y = R \end{cases}$$

Here S, I and R denote the densities of susceptible, infected and recovered populations, so that one has $S + I + R = 1$ at any time. We assume that all the parameters are known, and we aim at estimating S and I. Measuring R is equivalent to measuring $S + I$, and we rewrite this model as follows

$$
\begin{cases}
\dot{S} = -\beta I + \varphi(S, I) + \mu(1 - S - I) \\
\dot{I} = \beta I - \varphi(S, I) - \gamma I \\
\\
y = S + I
\end{cases}
$$

where we posit

$$
\varphi(S, I) = \beta(1 - S)I
$$

If we consider the initial stage of an epidemics, variables I and S are close to 0 and 1 respectively, that is on a domain $\Delta := \{1 - \varepsilon < S < 1; \ 0 < I < \varepsilon\}$ for some $\varepsilon > 0$. This domain is not invariant by the dynamics but we shall consider the estimation problem on a time windows for which the solution stays in this set Δ. One can check that the function φ is Lipschitz with a constant $\ell = 2\varepsilon$ on Δ, and can consider an extension of φ outside Δ with the same Lipschitz constant ℓ but on all $\mathbb{R}^2$, for instance

$$
\tilde{\varphi}(I, S) = \varphi(sat_{[0,\varepsilon]}sat_{[0,\varepsilon]}(I))
$$

where $sat_{[\,]}$ denotes the saturation function

$$
sat_{[a,b]}(x) = \max(a, \min(b, x)).
$$

On Δ, the system can be written as follows

$$
\frac{d}{dt}\begin{bmatrix} S \\ I \end{bmatrix} = \underbrace{\begin{bmatrix} -\mu & -\beta - \mu \\ 0 & \beta - \gamma \end{bmatrix}}_{A}\begin{bmatrix} S \\ I \end{bmatrix} + \underbrace{\begin{bmatrix} \tilde{\varphi}(S, I) \\ -\tilde{\varphi}(S, I) \end{bmatrix}}_{\phi(S,I)}
$$

$$
y = \underbrace{[1\ 1]}_{C}\begin{bmatrix} S \\ I \end{bmatrix}
$$

(4.14)

One can check that the pair (A, C) is observable, as the matrix

$$
\begin{bmatrix} C \\ CA \end{bmatrix} = \begin{bmatrix} 1 & 1 \\ -\mu & -\mu - \gamma \end{bmatrix}
$$

is full rank. The system (4.14) is well defined on the whole $\mathbb{R}^2$, and ϕ is a globally Lipschitz map, with a Lipschitz constant equal to 2ℓ. We are thus in position to apply Proposition 4.1: if ε is small enough, the system

$$\frac{d}{dt}\begin{bmatrix}\hat{S}\\\hat{I}\end{bmatrix} = A\begin{bmatrix}\hat{S}\\\hat{I}\end{bmatrix} + \phi(\hat{S}, \hat{I}) + G(\hat{S} + \hat{I} - y(t))$$

where G is such that $A + GC$ is Hurwitz, is an exponential observer of (4.14).
   In Sect. 5.2.2, we discuss the applicability of this observer with numerical values.

   It may happen that the estimation error of an observer is only partially assignable, as we shall see in the next example.

4.4 Observers via Decoupled Variables

Consider, as in Sect. 2.5, systems

$$\begin{cases} \dot{x} = f(x), & x \in \Omega \\ y = h(x) \end{cases}$$

for which there exists a change of coordinates $x \in \Omega \mapsto w = g(x, h(x)) \in \mathcal{W} \subset \mathbb{R}^{n'}$ such that one has

$$\dot{w}(t) = l(w(t), h(x(t))), \quad t \geq 0$$

for any solution $x(\cdot)$ in Ω, with the properties

1. $\{w = g(x, h(x)) \iff x = \tilde{g}(w, h(x))\}$, $(x, w) \in \Omega \times \mathcal{W}$
2. $\{w = g(x, h(x)) \implies h(x) = k(w)\}$, $(x, w) \in \Omega \times \mathcal{W}$

where $g, \tilde{g}, l$ and k are smooth maps.
   Then, one can look for an observer $\hat{w}(\cdot)$ of the system

$$\begin{cases} \dot{w} = l(w, h(x)) \\ y = k(w) \end{cases}$$

and take, as an estimator of $x(\cdot)$

$$\hat{x}(t) = \tilde{g}(\hat{w}(t), y(t)). \tag{4.15}$$

There is an advantage of considering such a change of coordinates when the maps $g, \tilde{g}, l$ and k are independent of a parameter θ present in the expression of f, as in Proposition 2.6 of Sect. 2.5. However, note that the estimator (4.15) does not filter the measurement $y(\cdot)$ and might be sensitive to noise.

Let us illustrate this approach on the malaria model (Example 2.6).

Example 4.3 Consider the model (2.14) of Example 2.6. With the variable $w = x - Ey = (I - EC)x$ in $\mathbb{R}^7$, the dynamics (2.15) is independent of the unknown non-linear term βSM:

$$
\begin{cases}
\dot{w} = \bar{A}w + \bar{A}Ey + \Lambda\,e_1 \\
\\
y = Cx
\end{cases}
$$

and one can consider the following observer for system (2.15) in Luenberger form

$$
\begin{cases}
\dot{\hat{w}} &= \bar{A}\hat{w} + \bar{A}Ey + \Lambda\,e_1 + L(y(t) - C(\hat{w} + Ey(t))) \\
&= (\bar{A} - LC)\,\hat{w} + (L + (\bar{A} - LC)E)\,y(t) + \Lambda\,e_1, \\
\\
\hat{x}(t) = \hat{w}(t) + E\,y(t).
\end{cases}
\tag{4.16}
$$

where L is a gains vector in $\mathbb{R}^7$ to be chosen. The dynamics of the error $e(t) = \hat{x}(t) - x(t)$ is given by

$$
\dot{e} = (\bar{A} - LC)e
$$

Note that one has $C\bar{A} = 0$. Therefore the rank of the observability matrix of the pair $(\bar{A}, C)$ is equal to one. The choice of L allows then to assign only one eigenvalue of $\bar{A} - LC$, equal to $-(L_2 + L_3)$, the other eigenvalues remaining negative. Therefore (4.16) is an observer for system (2.15) with exponential convergence, that does not use the unknown parameter β.

Remark 4.2 Differently to observers of Sect. 4.2, one cannot expect a convergence speed of the observer (4.16) faster than the dynamics (2.14), because the error dynamics is not completely assignable. However, the convergence is exponential. This is illustrated with numerical simulations in Sect. 5.2.4.

4.5 Reduced-Order Observers

A typical situation is when one can operate a state decomposition when $m < n$, as follows

1. decompose the state vector x (may be at the price of a change of variables) as

$$
x = \begin{bmatrix} y \\ x_u \end{bmatrix}
$$

where $x_u \in \mathbb{R}^{n-m}$ represent the *unmeasured* variables,

2. look for an auxiliary variable (say z) whose dynamics is independent of x_u

$$\dot{z} = g(z, y(t)), \ z \in \mathbb{R}^q$$

(for some q) and asymptotically stable (that is any solution $z(\cdot)$ converges to 0 when $t \rightarrow +\infty$), and such that x_u can be globally expressed as

$$x_u = l(z, y)$$

where l is a smooth map (say C^1).

Then, the dynamics

$$\begin{cases} \dot{\hat{z}} = g(\hat{z}, y(t)) \\ \hat{x}_u = l(\hat{z}, y(t)) \end{cases}$$

is an *asymptotic observer*, whose error convergence $\hat{x}_u - x_u$ is simply provided by the asymptotic convergence of $\hat{z}$ to 0, whatever is the initial condition $\hat{z}(0) \in \mathbb{R}^q$. When the convergence speed of an estimator cannot be adjusted, it is usually called an *asymptotic observer*, differently to the previous section for which the error convergence can be made arbitrarily fast. Note that differently to the previous section, these observers have no tuning parameters and are not driven by innovation terms. These estimators are *reduced-order* observers when the variable z is of lower dimension than x (i.e. when $q < n$). An interest for such observers is that it can possess robustness features when the maps g and l are independent of some terms or parameters of the dynamics $\dot{x} = f(x)$. Let us illustrate this feature on the Kermack-McKendrick model with fluctuating rates.

Example 4.4 We consider the SIR model with birth and death terms

$$\begin{cases} \dot{S} = -\beta(t)SI + \nu N - \mu S \\ \dot{I} = \beta(t)SI - \rho(t)I - \mu I \\ \dot{R} = \rho(t)I - \mu R \end{cases} \tag{4.17}$$

where parameters β and ρ fluctuate unpredictably over time. We assume, for simplicity, that the birth rate ν is equal to the death rate μ, so that the total population remains constant of size $N = S + I + R$ (assumed to be known). Let us suppose that the size of the infected population is monitored over time as well as the number of new cured individuals, which amounts considering that the observation vector at time t is

$$y(t) = \begin{bmatrix} y_1(t) \\ y_2(t) \end{bmatrix} = \begin{bmatrix} I(t) \\ \rho(t)I(t) \end{bmatrix}.$$

Stocks of classes S and R are not initially known. Then, the system

$$
\begin{cases}
\dot{Z} = \nu N - y_2(t) - \mu Z \\
\hat{S} = Z - y_1(t) \\
\hat{R} = N - Z
\end{cases}
\tag{4.18}
$$

is an observer allowing to estimate S and R without knowing $\beta(\cdot)$ and $\rho(\cdot)$. Indeed, the dynamics of the estimators verifies

$$
\frac{d}{dt}(\hat{S} - S) = \dot{Z} - \dot{y}_1 - \dot{S} = -\mu(Z - S - I) = -\mu(\hat{S} - S)
$$

$$
\frac{d}{dt}(\hat{R} - R) = -\dot{Z} - \dot{R} = -\nu N + \mu(Z + R) = -\mu(\hat{R} - R)
$$

which ensures the convergence of the $\hat{S}$ and $\hat{R}$ estimates. Note that the internal dynamics of the observer is here of smaller dimension than the system, and that the estimate of the unmeasured state variable S is a function of the internal state Z of the observer and the observation y_1. The speed of convergence of this observer is not adjustable, but it has the advantage of being perfectly robust to any (unknown) variations of the terms $\beta(\cdot)$ and $\rho(\cdot)$. This is illustrated with numerical simulations in Sect. 5.2.3.

4.6   The High-Gain Observer for Nonlinear Systems

In the two previous examples, the dynamics of the estimation error was linear. For an observable non-linear system, the existence of an observer whose estimation error is linear is not guaranteed. This is a difficult problem. However, one can consider the (nonlinear) *observability canonical form* in $\mathbb{R}^n$ [54] (given here for a scalar output i.e. for $m = 1$)

$$
\dot{z} = F(z) := \underbrace{\begin{bmatrix} 0 & 1 & 0 & \cdots & & \\ 0 & 0 & 1 & 0 & \cdots & \\ & & \ddots & \ddots & & \\ & & & \ddots & \ddots & \\ & & & & 0 & 1 \\ & & & & & 0 \end{bmatrix}}_{A} z + \psi(z) \underbrace{\begin{bmatrix} 0 \\ \vdots \\ 0 \\ 1 \end{bmatrix}}_{B}
\tag{4.19}
$$

$$
y = \underbrace{[1\,0\,\cdots\cdots\,0]}_{C} z
\tag{4.20}
$$

where the function ψ is Lipschitz on $\mathbb{R}^n$. Then, one can show that there exists an observer of the Luenberger form

$$\dot{\hat{z}} = F(\hat{z}) + G(C\hat{z} - y(t)) \tag{4.21}$$

with exponential convergence when G is a well-chosen gains vector. When an observable system

$$\begin{cases} \dot{x} = f(x), & x \in \mathbb{R}^n \\ y = h(x), & y \in \mathbb{R} \end{cases}$$

is not in normal form, but the application

$$\phi_n(x) = \begin{bmatrix} h(x) \\ \mathcal{L}_f h(x) \\ \vdots \\ \mathcal{L}_f^{n-1} h(x) \end{bmatrix}$$

is a diffeomorphism[2] from $\mathbb{R}^n$ into $\mathbb{R}^n$ and the function

$$\psi(z) := \mathcal{L}_f^m h \circ \phi_n^{-1}(z)$$

is Lipschitz on $\mathbb{R}^n$, then the observer (4.21) can be written in the x coordinates as follows

$$\dot{\hat{x}} = f(\hat{x}) + [J\phi_n(\hat{x})]^{-1} G(h(\hat{x}) - y(t))$$

where $J\phi_n(x)$ denotes the Jacobian matrix of ϕ at x. The observer preserves the Luenberger structure but with variable gains.

Let us first note that the pair (A, C) as defined in (4.19)–(4.20) is observable. Indeed, we have $O = Id$. Thus, according to the Lemma 4.1, one can freely assign the spectrum of $A + GC$ by choosing the vector G. We show now how to choose the eigenvalues of $A + GC$ to ensure the convergence of the non-linear observer (4.21). To do this, we begin by giving some properties of the Vandermonde matrices

$$V_{\lambda_1, \cdots, \lambda_n} := \begin{bmatrix} \lambda_1^{n-1} & \lambda_1^{n-2} & \cdots & \lambda_1 & 1 \\ \lambda_2^{n-1} & \lambda_2^{n-2} & \cdots & \lambda_2 & 1 \\ \vdots & \vdots & & \vdots & \vdots \\ \lambda_n^{n-1} & \lambda_n^{n-2} & \cdots & \lambda_n & 1 \end{bmatrix}$$

related to the normal form.

---

[2] A diffeomorphism is an invertible map such that both the map and its inverse are differentiable.

Lemma 4.2 *Let* $\Lambda = \{\lambda_1, \cdots, \lambda_n\}$ *be a set of n distinct real numbers and G a vector such that*

$$Sp(A + GC) = \Lambda := \{\lambda_1, \cdots, \lambda_n\}$$

Then

$$V_{\lambda_1, \cdots, \lambda_n}(A + GC)V_{\lambda_1, \cdots, \lambda_n}^{-1} = \begin{bmatrix} \lambda_1 & & & \\ & \lambda_2 & & \\ & & \ddots & \\ & & & \lambda_n \end{bmatrix}$$

Moreover, for any numbers $c > 0$ *and* $\theta > 0$, *there exist* $\lambda_n < \lambda_{n-1} < \cdots < \lambda_1 < 0$ *such that*

$$\lambda_1 + c||V_{\lambda_1(\theta), \cdots, \lambda_n(\theta)}^{-1}||_\infty = -\theta$$

The proof of Lemma 4.2 is given in Appendix A.2.

We are now ready to show the convergence of the observer (4.21) in coordinates z, for a gains vector G such that $A + GC$ has n distinct eigenvalues λ_1, ..., λ_n of negative real values. Denote the error $e = \hat{z} - z$. We have

$$\dot{e} = (A + GC)e + B(\psi(\hat{z}) - \psi(z))$$

Let $\xi = Ve$ where V designates the Vandermonde matrix $V_{\lambda_1, \cdots, \lambda_n}$. Thanks to Lemma 4.2, we obtain

$$\dot{\xi} = \Delta\xi + VB(\psi(\hat{z}) - \psi(z))$$

where Δ is the diagonal matrix $diag(\lambda_1, \cdots, \lambda_n)$. By multiplying on the left by $\xi^\top$, one obtains

$$\begin{aligned} \xi^\top\dot{\xi} &= \xi^\top\Delta\xi + \xi^\top VB(\psi(\hat{z}) - \psi(z)) \\ &\leq \lambda_1||\xi||^2 + ||\xi||\sqrt{n}|\psi(\hat{z}) - \psi(z)| \\ &\leq \lambda_1||\xi||^2 + ||\xi||\sqrt{n}L||e|| \\ &\leq (\lambda_1 + \sqrt{n}L||V^{-1}||_\infty)||\xi||^2 \end{aligned}$$

where L is the Lipschitz constant of ψ. Thus the norm of ξ verifies

$$||\xi(t)|| \leq ||\xi(0)|| + \int_0^t (\lambda_1 + \sqrt{n}L||V^{-1}||_\infty)||\xi(\tau)||d\tau$$

and by Gronwall's Lemma, we obtain

$$||\xi(t)|| \leq ||\xi(0)||e^{\left(\lambda_1 + \sqrt{n}L||V^{-1}||_\infty\right)t}$$

Finally, for any $\theta > 0$, Lemma 4.2 gives the existence of numbers $\lambda_n < \lambda_{n-1} < \cdots < \lambda_1 < 0$ such that $||\xi(t)|| \leq ||\xi(0)||e^{-\theta t}$ for any $t > 0$, which guarantees the exponential convergence of the error e to 0.

The observer (4.21) with the gains vector G_θ is called *high gain observer* [54], because the value of θ must be "sufficiently" large, and its successive powers might take large values.

Remark 4.3 In practice, the function ψ is not necessarily globally Lipschitz on $\mathbb{R}^n$, and even not properly defined outside Ω, while the observer (4.21) needs to defined on whole $\mathbb{R}^n$. Nevertheless, if there exists a compact subset K of Ω that is forwardly invariant by the dynamics (4.19), one can consider an extension of ψ outside K that is globally Lipschitz on $\mathbb{R}^n$ and define then the observer on whole $\mathbb{R}^n$ (see for instance [60, 102]). This is illustrated on Example 4.5 below. It could also happen that for observable systems the map ϕ_n is not injective (and thus cannot be a diffeomorphism), but $\phi_{\tilde{n}}$ is injective for $\tilde{n} > n$. Then, it is theoretically possible to embed the system in $\mathbb{R}^{\tilde{n}}$, that is to write the dynamics with a state vector in dimension $\tilde{n} > \tilde{n}$ and then to construct a high gain observers in $\mathbb{R}^{\tilde{n}}$. Such construction is beyond the present book (see [102] for some techniques to build such extensions).

Let us now illustrate this construction on the Kermack-McKendrick model.

Example 4.5 We consider the classical SIR model

$$\begin{cases} \dot{S} = -\beta SI \\ \dot{I} = \beta SI - \rho I \\ \dot{R} = \rho I \end{cases} \tag{4.22}$$

where the parameters β and ρ are known, and suppose that the only observation is the cumulative number of recovered individuals since a time t_0

$$y(t) = \int_{t_0}^{t} \rho I(\tau)d\tau \,.$$

It is also assumed that the size of the total population $N = S + I + R$ is known. To put the system in canonical form, we write

$$z_1 = y$$
$$z_2 = \dot{z}_1 = \rho I$$
$$z_3 = \dot{z}_2 = (\beta S - \rho)z_2$$

and

$$\dot{z}_3 = (\beta S - \rho)z_3 - \beta(\beta S I)z_2 = \psi(z) := \frac{z_3^2}{z_2} - \frac{\beta}{\rho}\left(\frac{z_3}{z_2} + \rho\right)z_2^2 \, .$$

We can then reconstruct the I, S and R stocks from the variables z as follows

$$S = \frac{1}{\beta}\left(\frac{z_3}{z_2} + \rho\right)$$

$$I = \frac{z_2}{\rho}$$

$$R = N - \frac{1}{\beta}\left(\frac{z_3}{z_2} + \rho\right) - \frac{z_2}{\rho}$$

Note that ψ is not globally Lipschitz on $\mathbb{R}^3$, and has a singularity at $z_2 = 0$. Nevertheless, we notice that the term z_3/z_2 can be framed as follows

$$\frac{z_3}{z_2} = \beta S - \rho \in [-\rho, \beta - \rho]$$

and that one has

$$\dot{z}_3 \in [-\rho^3 N - \rho\beta^2 N^3, \rho(\beta - \rho)N] \, .$$

We can therefore consider the expression

$$\tilde{\psi}(z) = sat_{[-\rho^3 N - \rho\beta^2 N^3, \rho(\beta-\rho)N]}\left(sat_{[-\rho,\beta-\rho]}\left(\frac{z_3}{z_2}\right)z_3 - \frac{\beta}{\rho}z_3 z_2 - \beta z_2^2\right)$$

instead of $\psi(z)$, where $sat_{[]}$ denotes the saturation function

$$sat_{[a,b]}(x) = \max(a, \min(b, x)) \, .$$

Finally, we choose the gains G_i of the observer such that $Sp(A + GC) = \{\lambda_1, \cdots, \lambda_n\}$ with $\lambda_n < \lambda_{n-1} < \cdots < \lambda_1 < 0$ and λ_1 *enough* negative. This amounts to take $G_i = \sigma_i(\{\lambda_1, \cdots, \lambda_n\})$, where $\sigma_i(\cdot)$ are the symmetric functions defined in (4.6). One thus obtains the internal dynamics of the observer

$$\begin{cases} \dot{\hat{z}}_1 = \hat{z}_2 + (\lambda_1 + \lambda_2 + \lambda_3)(\hat{z}_1 - y(t)) \\ \dot{\hat{z}}_2 = \hat{z}_3 + (\lambda_1\lambda_2 + \lambda_1\lambda_3 + \lambda_2\lambda_3)(\hat{z}_1 - y(t)) \\ \dot{\hat{z}}_3 = \tilde{\psi}(\hat{z}) + \lambda_1\lambda_2\lambda_3(\hat{z}_1 - y(t)) \end{cases} \qquad (4.23)$$

and the estimators

$$
\begin{cases}
\hat{S} = \dfrac{1}{\beta}\left(sat_{[-\rho,\beta-\rho]}\left(\dfrac{\hat{z}_3}{\hat{z}_2}\right) + \rho\right) \\[3mm]
\hat{I} = \dfrac{\hat{z}_2}{\rho} \\[3mm]
\hat{R} = N - \hat{S} - \hat{I}
\end{cases}
\tag{4.24}
$$

4.7   Discussion

The construction of an observer can avoid in certain situations to study the identifiability. For instance, in Sects. 4.4 and 4.5, the observers do not require the knowledge of all the parameters of the model, and even of some functions involved in the model. This is why such observers are also called *unknown-inputs observers*. The theory of unknown-inputs observers has been mainly studied for linear systems [32, 66, 90]. Very few general results are available for nonlinear systems (this research field is today largely open).

The existence of observers without the possibility of fixing arbitrarily the speed of convergence, as in Examples 4.3 and 4.4, is connected to the property of *detectability* (see for instance [9, 88]), which is a weaker property than observability: a system

$$
\begin{cases}
\dot{x} = f(x), \ x \in \Omega \\[2mm]
y = h(x)
\end{cases}
$$

is *detectable* (in Ω) if for any pair of solutions $x^a(\cdot)$, $x^b(\cdot)$ in Ω, one has

$$
\left\{ h(x^a(t)) = h(x^b(t)), \ t \geq 0 \right\} \ \Rightarrow \ \lim_{t\to+\infty} x^a(t) - x^b(t) = 0.
$$

For linear dynamics $\dot{x} = Ax$, $y = Cx$ that are not observable, there exists a *Kalman decomposition* [71] i.e. an invertible matrix P such that

$$
PAP^{-1} = \begin{bmatrix} A_{11} & 0 \\ A_{21} & A_{22} \end{bmatrix}, \quad CP^{-1} = \begin{bmatrix} C_1 & 0 \end{bmatrix}
$$

with $A_{11} \in \mathcal{M}_{l\times l}$, $C_1 \in \mathcal{M}_{1\times l}$ where $l < n$ is the rank of the observability matrix O recalled in (2.4), such that the subsystem $\dot{z} = A_{11}z$, $y = C_1 z$ is observable. Then, according to [71], the system is detectable when the matrix A_{22} is Hurwitz. This is exactly the case of Example 4.3 with $l = 1$.

Let us mention the more recent technique proposed by Kazantzis and Kravaris for obtaining Luenberger like observers for non-linear systems. It consists in looking at a (non-linear) change of coordinates as a diffeormorphism $T: x \mapsto \zeta = T(x)$ such that the dynamics of the variable ζ writes

$$\dot{\zeta} = A\zeta + B(h(T^{-1}(x)))$$

where A is a Hurwitz matrix and B a smooth map. Then,

$$\dot{z} = AZ + B(y), \quad \hat{x} = T^{-1}(z)$$

is a natural observer, whose speed of convergence is given by the spectrum of the matrix A. However, the map T has to be found as a solution of the partial derivative equation

$$\frac{\partial T}{\partial x}(x) f(x) = AT(x) + B(h(x))$$

and has to be Lipschitz and invertible for $\hat{x}$ to be well defined and converging. This is a difficult problem to solve (see [10, 75]), still open in general. This is why we have not presented this method.

Other approaches consider non-smooth observers. In particular, the following observer proposed by Levant [78]

$$\begin{cases} \dot{z}_1 = z_2 - K_1 L^{\frac{1}{3}} |z_1 - y|^{\frac{2}{3}} sign(z_1 - y) \\ \dot{z}_2 = z_3 - K_2 L^{\frac{1}{2}} |z_1 - y|^{\frac{1}{3}} sign(z_1 - y) \\ \dot{z}_3 = -K_3 L sign(z_1 - y) \end{cases}$$

reconstructs theoretically $\dot{y}$ and $\ddot{y}$ in finite time from the measurement y, provided that the solutions of the original system are bounded. However, such observers and more generally sliding-mode observers [114], are extremely sensitive to measurement noise. Although well employed for mechanical or electro-mechanical systems, we believe that there are not very well suited to epidemiological models.

Chapter 5
Practical and Numerical Considerations

Abstract The objective of this chapter is to show how to put theory in practice, illustrated in some cases studies.

5.1 Practical Identifiability

Till now we have studied structural identifiability/observability. While structural identifiability is a property of the model structure, given a set of outputs, practical identifiability is related to the actual data. In particular, it depends on the amount of information contained in the data.

A model can be structurally identifiable, but still be practically unidentifiable due to poor data quality, e.g., bad signal-to-noise ratio, errors in measurement or sparse sampling [103]. Structural identifiability means that parameters are identifiable with ideal (continuous, noise-free) data. While structural identifiability is a prerequisite for parameter identification, it does not guarantee that parameters are practically identifiable with a finite number of noisy data points.

Moreover, parameter estimation requires using numerical optimization algorithms. The distance, for the problem considered, to the nearest ill-posed problem, [43, 64], i.e., the conditioning of the problem, can challenge the convergence of algorithms.

Another source of practical unidentifiability is the lack of information from the data, i.e., the signal from the data does not satisfy the *persistence of excitation* [83]. This is the case when the observation is near an equilibrium [15].

In this section we use sensitivity analysis and results from asymptotic statistical theory to study practical identifiability. We refer to previous surveys and papers on the topic [13, 14, 16, 36, 41]. Our purpose here is to give an intuitive account of these techniques.

© The Author(s), under exclusive license to Springer Nature Singapore Pte Ltd. 2024 59
N. Cunniffe et al., *Identifiability and Observability in Epidemiological Models*,
SpringerBriefs on PDEs and Data Science,
https://doi.org/10.1007/978-981-97-2539-7_5

5.1.1   Rationale for Using Sensitivity Analysis

Practical identifiability is often assessed in terms of confidence intervals on parameters [128]. Confidence intervals can be derived from the Fisher Information Matrix (FIM) [23]. More specifically, the covariance matrix (Σ) of the estimated parameters may be approximated as the inverse of the FIM. The diagonal elements of $\Sigma \approx \text{FIM}^{-1}$ correspond to the variance of the parameter estimates. Their square-roots (the standard deviations) give confidence intervals on the parameters, thus providing information on practical identifiability.

In the least-squares framework, the Fisher Information Matrix can be expressed in terms of sensitivity matrices, that we define below.

5.1.2   Observed System

We consider that the initial condition x_0 is unknown. Unless otherwise specified, the term "parameter" now refers to both the parameter θ and the initial condition x_0, i.e. $\Theta = (\theta, x_0)$. We make explicit the dependence of the state variables x and y on Θ to clarify the following derivations:

$$
\begin{cases}
\dot{x}(t, \Theta) = f(x(t, \Theta), \theta), \quad x(0, \Theta) = x_0, \\
\dot{\theta}(t) = 0, \quad \theta(0) = \theta, \\
\\
y(t, \Theta) = h(x(t, \Theta), \theta).
\end{cases}
\tag{5.1}
$$

with $x \in \mathbb{R}^n$, $y \in \mathbb{R}^m$ and $\theta \in \mathbb{R}^p$.

5.1.3   Sensitivity Analysis

We wish to quantify how the observed variable $y(t, \Theta)$ changes for a small parameter variation $\Delta\Theta$.

We denote the Jacobian of the observation $y(t, \Theta)$ with respect to the parameter Θ as

$$
\chi(t, \Theta) = \frac{\partial y}{\partial \Theta}(t, \Theta).
$$

This $m \times (n + p)$ matrix is called the sensitivity matrix.

By linearization (first-order Taylor approximation), one can write

$$\Delta y(t, \Theta) = \chi(t, \Theta)\,\Delta\Theta.$$

Let us give some remarks.

Reid [104] defined a parameter vector as "sensitivity identifiable" if the above equation can be solved uniquely for $\Delta\Theta$. This linear problem is well known: if χ has maximal rank then the solution is given by means of the Moore-Penrose pseudo-inverse $\chi^+ = (\chi^T\chi)^{-1}\chi^T$:

$$\Delta\Theta = \chi^+(t, \Theta)\,\Delta y(t, \Theta).$$

It is also well known [55] that the sensitivity of this solution is ruled by the condition number $\kappa_2(\chi) = \sigma_{max}/\sigma_{min}$ with σ_{max} and σ_{min} respectively the greatest and smallest singular value of χ (which are the square roots of the corresponding eigenvalues of $\chi^T\chi$).

5.1.4 Ordinary Least Squares

Now we consider a set of M observations Y_i, $i = 1, \ldots, M$, that have been obtained at times t_i. We assume that the observation is given by

$$Y_i = y(t_i, \Theta) + \mathcal{E}_i,$$

with the error $\mathcal{E}_i$ assumed to be a random variable satisfying the following assumptions:

- the errors $\mathcal{E}_i$ have mean zero $E[\mathcal{E}_i] = 0$;
- the errors have a constant variance $\mathrm{var}(\mathcal{E}_i) = \sigma^2$;
- the errors are independent and identically distributed.

The Fisher Information Matrix, for the preceding defined observations, is defined as

$$\mathrm{FIM}(\Theta, \sigma^2) = \frac{1}{\sigma^2}\sum_{i=1}^{M} \chi(t_i, \Theta)^T\,\chi(t_i, \Theta). \tag{5.2}$$

Solving the ordinary least square (OLS) equations gives an estimator $\hat{\Theta}_{OLS}$ of the parameter Θ:

$$\hat{\Theta}_{OLS} = \arg\min_{\Theta}\sum_{i=1}^{M} (Y_i - y(t_i, \Theta))^2. \tag{5.3}$$

Even though the error's distribution is not specified, asymptotic statistical theory can be used to approximate the mean and variance of the estimated Θ (a random variable) [16, 109]: the bias-adjusted approximation for σ^2 (with $n + p$ "parameters") is

$$\hat{\sigma}^2_{\text{OLS}} = \frac{1}{M - (n + p)} \sum_{i=1}^{M} \left(Y_i - y \left(t_i, \hat{\Theta}_{\text{OLS}} \right) \right)^2 . \tag{5.4}$$

5.1.5 Confidence Intervals

The above approximation of the error variance can be used to further approximate the parameter covariance matrix Σ:

$$\hat{\Sigma} := \left[\text{FIM} \left(\hat{\Theta}, \hat{\sigma}^2_{\text{OLS}} \right) \right]^{-1} . \tag{5.5}$$

The standard error (SE) for $\hat{\Theta}_{\text{OLS}}$ can be approximated by taking the square roots of the diagonal elements of the covariance matrix Σ: for all $k = 1, \ldots, n + p$,

$$\text{SE} \left(\hat{\Theta}_{\text{OLS}}(k) \right) = \sqrt{\hat{\Sigma}_{kk}} . \tag{5.6}$$

Finally, to compute the 95% confidence interval for the k-th component of the parameter vector $\hat{\Theta}_{\text{OLS}}$ with $n + p$ "parameters", one may use the Student's t–distribution with $M - (n + p)$ degrees of freedom: letting

$$\zeta(k) = t_{0.025}^{M-(n+p)} \times \text{SE} \left(\hat{\Theta}_{\text{OLS}}(k) \right) ,$$

the confidence interval is defined as

$$\hat{\Theta}_{\text{OLS}}(k) - \zeta(k) < \hat{\Theta}_{\text{OLS}}(k) < \hat{\Theta}_{\text{OLS}}(k) + \zeta(k) .$$

From these formulas it appears that the conditioning of the Fisher Information Matrix plays an essential role. Huge confidence intervals give indications about the practicality of the identification.

5.1.6 Computing the Sensitivity Matrix

The sensitivity matrix $\chi(t, \Theta)$, with $\Theta = (\theta, x_0)$, is obtained by integrating an ODE. The components of the ODE to be integrated depend on whether one differentiates with respect to θ or x_0.

1. We first differentiate the output function with respect to θ.
   The first part of the ODE is given by

$$\frac{\partial y}{\partial \theta}(t, \theta, x_0) = \frac{d}{d\theta} h(x(t, \theta, x_0), \theta) = \frac{\partial h}{\partial x} \frac{\partial x}{\partial \theta}(t, x_0, \theta) + \frac{\partial h}{\partial \theta}(x(t, \theta, x_0), \theta).$$

The Jacobian $\dfrac{\partial h}{\partial x}$ is a $m \times n$ matrix while $\dfrac{\partial h}{\partial \theta}$ is a $m \times p$ matrix.
We then have to compute the $n \times p$ matrix

$$z(t, \Theta) := \frac{\partial x}{\partial \theta}(t, \theta, x_0) = \frac{\partial x}{\partial \theta}(t, \Theta).$$

Let $A(t, \Theta)$ and $B(t, \Theta)$ be the following time-dependent $n \times n$ and $n \times p$ matrices, respectively:

$$A(t, \Theta) := \frac{\partial f}{\partial x}(x(t, \Theta), \theta),$$

and

$$B(t, \Theta) := \frac{\partial f}{\partial \theta}(x(t, \Theta), \theta).$$

It is well known [61] that $z(t)$ is the solution of the linear matrix equation:

$$\dot{z}(t, \Theta) = A(t, \Theta) z(t, \Theta) + B(t, \Theta),$$

with the initial condition $z(0, \Theta) = 0_{n \times p}$ (a zero matrix of size $n \times p$).
2. We differentiate now with respect to x_0.
   The second part of the ODE is given by

$$\frac{\partial y}{\partial x_0}(t, \theta, x_0) = \frac{\partial}{\partial x_0} h(x(t, \theta, x_0), \theta) = \frac{\partial h}{\partial x} \frac{\partial x}{\partial x_0}(t, \theta, x_0).$$

Let

$$w(t, \Theta) := \frac{\partial x}{\partial x_0}(t, \theta, x_0) = \frac{\partial x}{\partial x_0}(t, \Theta).$$

Based on the same reference [61], $w(t, \Theta)$ is solution of the linear matrix ODE

$$\dot{w}(t, \Theta) = A(t, \Theta)\, w(t, \Theta),$$

with the initial condition $w(0, \Theta) = \mathrm{Id}_{n \times n}$ (the identity matrix of size n).

3. We write now the full system.

To summarize, one has to solve the following system in dimension $n^2 + np + n$

$$\begin{cases} \dot{x}(t, \Theta) = f(x(t, \Theta), \theta), & x(0, \Theta) = x_0, \\ \dot{z}(t, \Theta) = A(t, \Theta)\, z(t, \Theta) + B(t, \Theta), & z(0, \Theta) = 0_{n \times p}, \\ \dot{w}(t, \Theta) = A(t, \Theta)\, w(t, \Theta), & w(0, \Theta) = \mathrm{Id}_{n \times n} \end{cases} \quad (5.7)$$

with $A(t, \Theta) = \dfrac{\partial f}{\partial x}(x(t, \Theta), \theta)$ and $B(t, \Theta) = \dfrac{\partial f}{\partial \theta}(x(t, \Theta), \theta)$.

For large systems, the computation of the different Jacobians can be prohibitive, in this case automatic differentiation software has to be used.

5.1.7   Some Case Studies

In this section, we consider two classical examples as case studies. These examples have been used in many books of mathematical epidemiology, e.g. [89].

Case 1. Influenza in a Boarding School

Our first example is an outbreak of influenza in a United Kingdom boarding school which occurred in 1978 [31]. In [89] the parameters β, γ are identified by an unspecified "best-fit" algorithm. A more complete analysis is done in [24], where the analysis is done using sensitivity analysis and asymptotic statistical theory. In [86] the same example is considered. Different sources exist for the data [24, 25, 42] with small differences.

Using the figure in [31] and the *Plot Digitizer* software, we got an approximation of the data. It was reported that $N = 763$, and the conditions at the start of this outbreak were $S_0 = 762$ and $I_0 = 1$. We used the following data, in which time t is in days and $I(t)$ denotes the number of infectious people at time t.

t	0	1	2	3	4	5	6	7	8	9	10	11	12	13
$I(t)$	1	6	26	73	222	293	258	237	191	124	68	26	10	3

Fig. 5.1 Boarding school
example

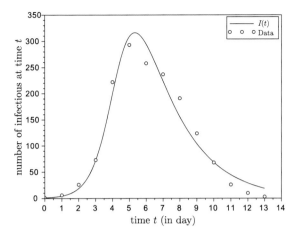

Specifically, we considered model (3.4) with $k = 1$ (all infectious are assumed to be observed). We obtained the following OLS estimation, as given by the Scilab software:

$$\beta \approx 1.96, \quad \gamma \approx 0.475$$

(see the numerical code in Appendix B). The fit is shown in Fig. 5.1.

We computed the 95% confidence intervals using the formulas given in the preceding section: (5.2), (5.3), (5.4), (5.5), (5.6), up to a few changes due to the fact that the initial conditions $x_0 = (S_0, I_0)$ are assumed to be known in this example (see Appendix B). We find:

$$\beta \approx 1.96 \pm 0.073,$$

$$\gamma \approx 0.475 \pm 0.04.$$

One can obtain approximately the same results using the likelihood profile method to compute confidence intervals [23], assuming normally distributed errors. However, it is well known that the profile method quickly becomes impractical for models with more than two parameters [23], which is the rule rather than the exception, as will be the case in the following example. This is why we stick to the FIM method. Lastly, we note that the condition number of the FIM is approximately equal to 3.80 in this example.

Case 2. Plague in Bombay
Our second example is the Bombay Plague of 1905–1906 [73]. We collected the data from [39, Table IX], over the same period as [73] (Dec. 17 to Jul. 21). The form of the data is presented in the following table, in which time t is in weeks, and $\dot{R}(t)$ denotes the number of deaths per week at time t.

t	0	1	2	3	4	5	6	$\cdots$	24	25	26	27	28	29	30
$\dot{R}(t)$	8	10	12	16	24	48	51	$\cdots$	106	64	46	35	27	28	24

We consider that the number of deaths per week is the same as $\dot{r}(t) = \gamma I(t)$, meaning that all infections lead to death, which is a reasonable assumption in this context [12]. Therefore, we consider model (3.4) with $k = \gamma$. In this example, not only the parameters β and γ, but also the size of the population, N, as well as the initial conditions, S_0 and I_0, are unknown [12]. According to Theorem 3.1 (and Remark 3.3), the model is neither observable nor identifiable. However, the model is partly identifiable in the sense that S_0, I_0, γ and $\tilde{\beta} = \beta/N$ are structurally identifiable. Starting from an arbitrary initial guess, we obtained the following OLS estimation, as given by the `Scilab` software:

$$\tilde{\beta} \approx 0.0000855, \quad \gamma \approx 3.72, \quad S_0 \approx 4.81 \ 10^4, \quad I_0 \approx 1.42,$$

(see the numerical code in Appendix C). The fit is shown in Fig. 5.2.

Proceeding as in the previous example, we obtained the following 95%-confidence intervals:

$$\tilde{\beta} \approx 0.0000855 \pm 0.00157,$$

$$\gamma \approx 3.72 \pm 25,$$

$$S_0 \approx 4.81 \ 10^4 \pm 6 \ 10^4,$$

$$I_0 \approx 1.42 \pm 34.$$

Fig. 5.2 Plague in Bombay example

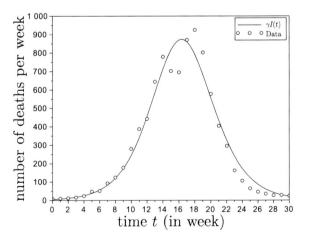

The confidence intervals are huge, which means that we can have absolutely no confidence in the estimated values of the parameters, even though the fit looks good and these parameters are structurally identifiable in principle. In practice, one can show that many other and very different combinations of the parameters can yield approximately the same fit. Note that if we did as if the initial conditions were known, the confidence intervals on $\tilde{\beta}$ and γ would be reasonable, as in the previous example. In this example, the condition number of the FIM is approximately 9.14×10^{24}, meaning that the problem is "sloppy" [34]; note however that while it is usual that a model is both sloppy and practically non identifiable, this is not always the case [38]. Altogether, we can conclude that there is a severe practical identifiability issue in this classical example.

5.1.8 Discussion

The SIR model of Kermack-McKendrick has been studied in a series of papers [28, 36, 37] where the problem of observability/identifiability is approached from the statistical point view: addressing parameter identifiability by exploiting properties of both the sensitivity matrix and uncertainty quantification in the form of standard errors. In this series of papers, structural observability and identifiability were not explicitly addressed. For example in [37] the authors identify $(S_0, I_0, \beta/N, \gamma)$ based on incidence observations—akin to Eq. (3.5) with $k = 1$—which we have proved to be structurally identifiable (see Proposition 3.2 and Remark 3.5). Similarly in [28] the authors seek to identify $(S_0, I_0, \beta, \gamma)$ which, with N known, are structurally identifiable in principle. However the authors encounter practical identifiability issues. This is a typical example of strictly practical unidentifiability (as in our second example, the Plague in Bombay).

Although a structural observability and identifiability analysis should be done as a prerequisite to a practical identifiability analysis, it does not suffice. Moreover, when doing practical identifiability analyses, the error structure of the data should be considered. For instance, sensitivity analyses can be extended to non-constant error variance through Generalized Least Squares (GLS), which makes it possible to test different ways of weighting errors [37]. Appendix D shows the principle of GLS and how to compute the Fisher Information Matrix in this case. We also provide an online example[1] based on the "Influenza in a Boarding School" data (since the "Plague in Bombay" example generates numerical optimization issues related to its practical non-identifiability).

An additional issue may occur when the output signal is not sufficiently informative (i.e., not *persistently exciting* [83]). For example when the data correspond to states near unobservability, e.g., near an equilibrium. In those cases, one has to wait to have data sufficiently far from equilibrium.

---

[1] https://github.com/nikcunniffe/Identifiability.

To conclude, the problem of observability and identifiability, either structural or practical, is far from being simple, even in relatively simple SIR models with seemingly good quality data [37]. Of course, the more complex the model, the more parameters there are to identify, the more serious the problem of identifiability.

5.2 Observers in Practice

In this section, we show how the various observers presented in Sect. 4 behave in practice, and the role of the tuning parameters. Up to now, we have assumed the measurements to be perfect i.e. not tainted with any noise. Since integration has good "averaging" properties, an observer is expected to filter noise or inaccuracies in the measurements. However, we will see that the filtering capacity of an observer is related to his convergence speed, which often leads to a "precision-speed" dilemma in the choice of the observer or his settings.

Let us underline that when identifiability/observability cannot be proved theoretically or is too difficult to be proven analytically, one can still look for an observer and study its asymptotic convergence, theoretically or numerically.

5.2.1 Observers with Linear Assignable Error Dynamics

We illustrate the observer (4.8) of the age-structured model (4.7) on simulations, for the following values of the parameters.

a_1	a_2	m_1	m_2	m_3	k	$\bar{r}_{min}$	$\bar{r}_{max}$
0.1	0.1	0.05	0.07	0.07	1.0	0.9	1.1

The following code is used to compute the gain vector G for a set of desired eigenvalues.

```
a1=0.1;a2=0.1;m1=0.05;m2=0.07;m3=0.07;
Sp=0.3*[-0.3,-0.33,-0.36];
A=[-a1-m1,0,0;
a1,-a2-m2,0;
0,a2,-m3];
C=[0,0,1];
B=[0;0;1];
Obs=[C;C*A;C*A*A];
L=inv(Obs)*B;
P=[L,A*L,A*A*L];
Abar=inv(P)*A*P;
sigma=coeff(poly(Sp,'x'));
G=P*(-sigma(1:3)'-Abar(:,3));
```

Figure 5.3 shows convergence for a moderately negative spectrum, while Fig. 5.4 shows the acceleration of convergence obtained for a spectrum located further to the left in the complex plane. For the same choice of gains, Figs. 5.5 and 5.6 show the effect of noise on the $y(\cdot)$ measurements. It can be seen that a faster convergence is more sensitive to noise and loses accuracy. In practice, one often has to make a compromise for the choice of the observer's setting.

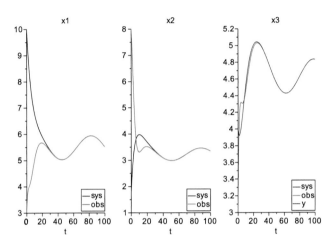

Fig. 5.3 $Sp(A + GC) = \{-0.3, -0.33, -0.36\}$ without measurement noise

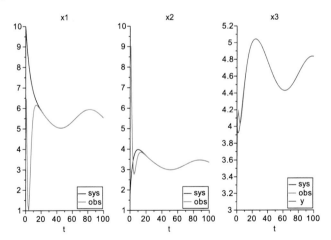

Fig. 5.4 $Sp(A + GC) = \{-0.6, -0.66, -0.72\}$ without measurement noise

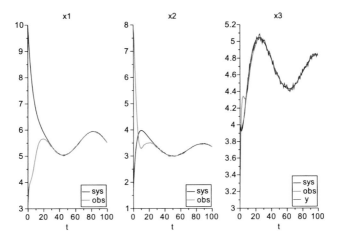

Fig. 5.5 $Sp(A + GC) = \{-0.3, -0.33, -0.36\}$ with measurement noise

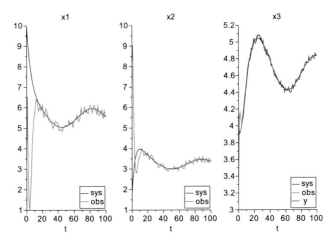

Fig. 5.6 $Sp(A + GC) = \{-0.6, -0.66, -0.72\}$ with measurement noise

5.2.2 About Observers with Lipchitz Non-linearity

Let us illustrate the result on Example 4.2 with the following numerical values of
the parameters

β	γ	μ
0.13	0.1	0.05

The gain vector G such that $Sp(A + GC) = \{-1, -2\}$, and the corresponding matrix P solution of the Lyapunov equation (4.11) can be numerically computed (using, for instance, `scilab`) as

```
G  = |  -23.889 |       P  = |  73.135667    83.710112 |
     |   20.909 |            |  83.710112    96.199374 |
```

with

```
    norm(P)  = 169.16821
```

for which the condition (4.12) gives $\varepsilon \leq 0.001477$. As an illustration, for a total population size of 1 billion, this gives $I \leq 1477$, which is a very small number...

This example shows that this technique is not well suited to epidemiological models such as the SIRS one, because it requires a too small Lipschitz constant of the non linear terms. However, we consider useful to have exposed this known approach and shown its drawback.

5.2.3 Observers with Asymptotic Convergence

We illustrate on simulations the behavior of the asymptotic observer (4.18) of the SIR model with fluctuating rates (4.17), for the following values of the parameters.

β	ρ	ν	μ	N
0.4 ± 0.08	0.2 ± 0.04	0.05	0.05	1000

Here β and ρ are functions of time chosen randomly in between the bounds given in the table. Figures 5.7 and 5.8 show that the observer has a convergence relatively insensitive to measurement noise, but the speed of convergence is slow because the exponential decay of the error is equal to μ, which is not adjustable.

Unlike the observers in previous sections, let us underline that the present observer is not based on a $\hat{y} - y$ innovation. Therefore, one is not informed of the quality of the estimate over time, which is a *price to pay* to have an observer insensitive to unknown variations of the epidemic parameters β, ρ.

5.2.4 Observers with Partially Assignable Error Dynamics

The observer given in (4.16) for the intra-host malaria model (2.14) is illustrated here on real data, as one can see for instance on Fig. 5.9.

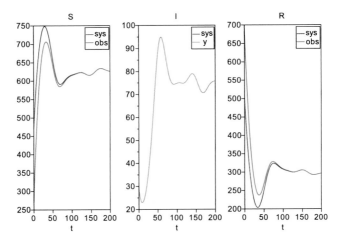

Fig. 5.7   The high gain observer without measurement noise

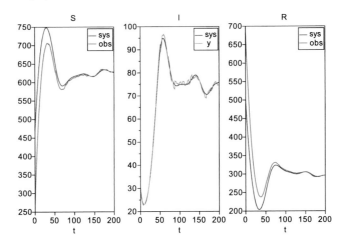

Fig. 5.8   The high gain observer with measurement noise

As already mentioned in Remark 4.2, the speed of convergence of this observer cannot be tuned as fast as desired. However, this is quite satisfactory in practice. Let us also underline that the observer does not require the reconstruction of the parameter β, although this parameter is identifiable (see Sect. 2.5). This is a strength of this observer, because the parameter β could switch or fluctuate with time.

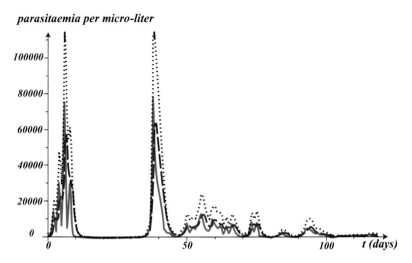

Fig. 5.9 Example of patient S1204: Measures (data) of peripheral parasitaemia are plotted with red solid line. The estimations delivered by the observer (4.16) are plotted with blue dashed line for the estimated sequestered parasitaemia, and with black dotted line for the estimated total parasitaemia. The gain used is $L = (0, 5, 5, 0, 0, 0, 0)^T$. $Sp(\bar{A}-LC) = \{-10, -\mu_S, -\mu_M, -\mu_5 - \gamma_5, -\mu_4 - \gamma_4, -\mu_3 - \gamma_3, -\mu_2 - \gamma_2 - \gamma_1\}$

5.2.5 High Gain Observer

The non-linear observer (4.23)–(4.23) of the classical SIR model (4.22) is illustrated on simulations for the following values

β	ρ	N
0.4	0.1	10000

where the y cumulative measures were made discretely every day (rounded to the nearest integer). In order to obtain a time-continuous $y(\cdot)$ signal, we performed an interpolation by cubic splines. Figure 5.10 shows the convergence of the observer for the eigenvalues $\{-2, -2.2, -2.4\}$. We also simulated the observer when the measurements are corrupted by random counting errors up to ± 5 individuals per day (see Fig. 5.11). As for the adjustable observer in Sect. 5.2.1, these simulations show the dilemma *accuracy versus speed* of the estimation in presence of measurement noise.

We end this section by showing an example for which an observer is used to reconstruct state and parameter simultaneously.

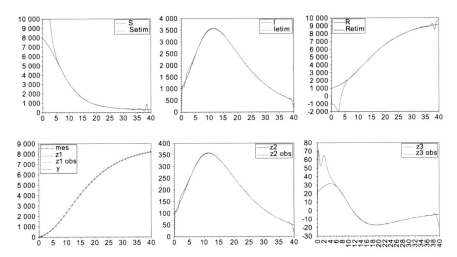

Fig. 5.10 Observer simulations without noisy measurements (top: variables S, I, R and their estimates; bottom: coordinates z_i with measurements points on the left)

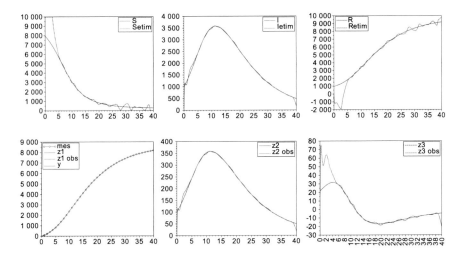

Fig. 5.11 Observer simulations with noisy measurements

5.3   A Case Study : An Observer to Estimate State and Parameter

Consider the following simple schistosomiasis (bilharzia) transmission model [17]:

$$\begin{cases} \dot{w} = -\gamma w + as, \\ \dot{s} = b(1 - s)w - \mu s, \end{cases} \tag{5.8}$$

where w is the number of female schistosomes (worms) per single host and s the proportion of infected snails. The female schistosomes *per* single host decays at a per capita rate γ and are replenished at a rate a. The latter process is proportional to the proportion of infected snails s. The proportion of susceptible snails $1 - s$ are infected through an indirect contact with schistosomes that are excreted from hosts w at rate b, and naturally die at a rate μ.

The number of female schistosomes (worms) *per* single host can be measured using urine or faeces samples. Therefore, we can assume that the measurable output is $y(t) = w(t)$.

It is easy to show that the compact set

$$\Omega = \{(w, s) \in \mathbb{R}^2 \mid 0 \le w \le \frac{a}{\gamma};\ 0 \le s \le 1\} \tag{5.9}$$

is a positively invariant set for System (5.8). The measurable output $y(t)$ satisfies the equation $\dot{y} = -\gamma y + as$. Hence, $y(t) > 0$ for all $t \ge 0$ if $y(0) > 0$.

We assume that the ecosystem is in an endemic situation which means that the basic reproduction number $\mathcal{R}_0 = \dfrac{ab}{\gamma\mu} > 1$ and implies [17] that $y(t)$ converges to $\bar{y} = \dfrac{a}{\gamma}\left(1 - \dfrac{\gamma\mu}{ab}\right) > 0$ as time goes to infinity.

A crucial problem in epidemic models is the estimation of the transmission parameters. For the schistosomiasis model, it is the parameter a which represents the snail-host infection rate that is difficult to estimate [17].

In order to study the observability and identifiability of model (5.8), we consider the augmented system by adding the unkown parameter a to the augmented state $x = (w, s, a)^\top$:

$$\begin{cases} \dot{x} = \begin{bmatrix} \dot{w} \\ \dot{s} \\ \dot{a} \end{bmatrix} = \begin{bmatrix} -\gamma w + as \\ b(1 - s)w - \mu s \\ 0 \end{bmatrix} = f(x) \\[2em] y = w = h(x) \end{cases} \tag{5.10}$$

One determines

$$\dot{y} = \mathcal{L}_f h(x) = \langle \nabla h(x) | f(x) \rangle = -\gamma w + as,$$

$$\ddot{y} = \mathcal{L}_f^2 h(x) = \langle \nabla \mathcal{L}_f h(x) | f(x) \rangle = \langle \begin{bmatrix} -\gamma \\ a \\ s \end{bmatrix} \Big| \begin{bmatrix} -\gamma w + as \\ b(1 - s)w - \mu s \\ 0 \end{bmatrix} \rangle$$

$$= -\gamma(-\gamma w + as) + a(b(1 - s)w - \mu s).$$

The map

$$H_3 : \{0 < w \le \frac{a}{\gamma};\ 0 < s < 1;\ a > 0\} \to \mathbb{R}^3$$

$$x \mapsto H_3(x) = \left(h(x), \mathcal{L}_f h(x), \mathcal{L}_f^2 h(x) \right)$$

is injective. Indeed

$$H_3(\bar{x}) = H_3(x) \Rightarrow \begin{cases} \bar{w} = w, \\ -\gamma \bar{w} + \bar{a}\bar{s} = -\gamma w + as, \\ \bar{a}(b(1 - \bar{s})\bar{w} - \mu\bar{s}) = a(b(1 - s)w - \mu s) \end{cases}$$

$$\Rightarrow \begin{cases} \bar{w} = w, \\ \bar{a}\bar{s} = as, \\ \bar{a}b(1 - \bar{s})\bar{w} = ab(1 - s)w \end{cases}$$

$$\Rightarrow \begin{cases} \bar{w} = w, \\ \bar{a}\bar{s} = as, \\ \bar{a} = a\ (\text{ since } b > 0, w > 0, \bar{w} > 0, s < 1, \bar{s} < 1) \end{cases}$$

$$\Rightarrow \begin{cases} \bar{w} = w, \\ \bar{s} = s\ (\text{ since } a > 0) \Rightarrow \bar{x} = x, \\ \bar{a} = a. \end{cases}$$

Therefore, thanks to Proposition 2.2, the augmented system (5.10) is observable. Using Chap. 1 Proposition 1.1, we deduce that Model (5.8) is observable and identifiable. Moreover, the state variables w and s as well as the parameter a can be expressed as rational functions of y, $\dot{y}$ and $\ddot{y}$ as follows:

$$\begin{cases} w = y, \\ s = \dfrac{(\gamma y + \dot{y})\, b}{(b + \mu)\gamma y + (b + \gamma + \mu)\dot{y} + \ddot{y}}, \\ a = \dfrac{(b + \mu)\gamma y + (b + \gamma + \mu)\dot{y} + \ddot{y}}{b} = \dfrac{\gamma y + \dot{y}}{s}. \end{cases}$$

Now, we shall build an observer that will allow to estimate the unmeasured state variable (here it is $s(t)$) as well as the unknown parameter a. To this end, we perform the following change of coordinates:

$$z_1 = w, \quad z_2 = -\gamma w + sa, \quad z_3 = a.$$

One has then

$$
\begin{cases}
\dot{z}_1 = -\gamma z_1 + z_3 s = -\gamma z_1 + z_3 \dfrac{z_2 + \gamma z_1}{z_3} = z_2, \\[2mm]
\dot{z}_2 = -\gamma(-\gamma z_1 + z_3 s) + (b(1 - s)w - \mu s)z_3 = -\gamma z_2 + (b(1 - s)w - \mu s)z_3 \\[2mm]
\quad\; = -\gamma z_2 + (b(1 - \dfrac{z_2 + \gamma z_1}{z_3})z_1 - \mu \dfrac{z_2 + \gamma z_1}{z_3})z_3 \\[2mm]
\quad\; = -\gamma z_2 + b(z_3 - z_2 - \gamma z_1)z_1 - \mu z_2 - \mu \gamma z_1, \\[2mm]
\dot{z}_3 = 0.
\end{cases}
$$

Note that, since $y = z_1$, the dynamics of z_2 can be written as

$$
\dot{z}_2 = -\gamma(\mu + by)z_1 - (\mu + \gamma + by)z_2. + byz_3
$$

Therefore, the dynamics takes the form

$$
\begin{cases}
\dot{z}(t) = A(y)\, z(t), \\[2mm]
y = C_0\, z,
\end{cases}
\tag{5.11}
$$

where

$$
A(y) = \begin{bmatrix} 0 & 1 & 0 \\ -\gamma(\mu + by) & -(\mu + \gamma + by) & by \\ 0 & 0 & 0 \end{bmatrix}, \quad C_0 = [1\ 0\ 0].
$$

For any fixed $y > 0$, the corresponding observability matrix is

$$
O_{(C_0, A)} = \begin{bmatrix} C_0 \\ C_0 A \\ C_0 A^2 \end{bmatrix} = \begin{bmatrix} 1 & 0 & 0 \\ 0 & 1 & 0 \\ -\gamma(by + \mu) & -by - \gamma - \mu & by \end{bmatrix}
$$

that is of full rank if $y \neq 0$. Therefore, by the pole-shifting theorem (see [129, page 61]), it is possible to find a y−dependent gain $K(y)$ such that $Sp(A(y) - K(y)C_0) = \{-\lambda_1, -\lambda_2, -\lambda_3\}$, where $sp(M(y))$ denotes the spectrum of $M(y)$ and λ_i are any positive real numbers. This gain $K(y)$ can be computed using for instance Ackermann's formula (see [11] page 382):

$$
K(y) = \prod_{i=1}^{3} (A(y) + \lambda_i I_3).O_{(C_0, A)}^{-1}. \begin{bmatrix} 0 \\ 0 \\ 1 \end{bmatrix},
$$

where I_3 is the 3×3 identity matrix.

$$K(y) = \begin{bmatrix} K_1(y) \\ K_2(y) \\ K_3(y) \end{bmatrix}$$

$$= \begin{bmatrix} -(\gamma + \mu) + \lambda_1 + \lambda_2 + \lambda_3 - by \\[2mm] \gamma^2 + \mu^2 + \mu\gamma - (\mu + \gamma)(\lambda_1 + \lambda_2 + \lambda_3) + (\lambda_2 + \lambda_3)\lambda_1 + \lambda_2\lambda_3 \\ +b^2 y^2 - (\lambda_1 + \lambda_2 + \lambda_3 - \gamma - 2\mu)by \\[2mm] \dfrac{\lambda_1\lambda_2\lambda_3}{by} \end{bmatrix}$$

The gain $K(y)$ is well defined since $y(t) > 0$ for all $t \geq 0$. An observer is then given by:

$$\dot{\hat{z}}(t) = A(y)\hat{z}(t) - K(y(t))(C_0\hat{z}(t) - y(t)). \tag{5.12}$$

in coordinates:

$$\begin{cases} \dot{\hat{z}}_1 = \hat{z}_2 - (\hat{z}_1 - y)K_1(y), \\ \dot{\hat{z}}_2 = -\mu\gamma\hat{z}_1 - (\mu + \gamma)\hat{z}_2 + b(\hat{z}_3 - \hat{z}_2 - \gamma\hat{z}_1)y - (\hat{z}_1 - y)K_2(y), \tag{5.13} \\ \dot{\hat{z}}_3 = -(\hat{z}_1 - y)K_3(y). \end{cases}$$

The error equation is

$$\dot{e}(t) = \Big(A(y) - K(y)\,C_0\Big)e(t) = M(y)\,e(t). \tag{5.14}$$

The eigenvalues of the matrix $M(y)$ are $-\lambda_1$, $-\lambda_2$ and $-\lambda_3$. It has been proved in [21] that the error e converges exponentially fast to zero (the proof being quite long is omitted here), which gives the exponential convergence of the observer (5.12).

β	μ	a	b
0.05	0.04	2	0.01

Finally, let us illustrate this observer on numerical simulations. We have taken the following values of the parameters The initial conditions are $(w(0), s(0)) = (3, 0.3)$ and $(\hat{z}_1(0), \hat{z}_2(0), \hat{z}_3(0)) = (1, \ 0.1, \ 0.2)$. The set of eigenvalues of the matrix $M(y)$ have been chosen to be $\{-0.4, -1.4, -2.4\}$. Figure 5.12 shows the convergence of the estimation of the unmeasured proportion of infected snails $s(t)$

$$\hat{s}(t) = \frac{\hat{z}_2(t) + \gamma \hat{z}_1(t)}{\hat{z}_3(t)}$$

and of unknown parameter a

$$\hat{a}(t) = \hat{z}_3(t)$$

delivered by the observer (5.13).

The same simulations have been conducted in `Scilab` with measurement noise

```
y=x(5)+0.1*grand(1,1,"nor",1,1)
```

Figure 5.13 shows that for this choice of gains, the estimations are heckled but follow quite well the unknown state and parameter.

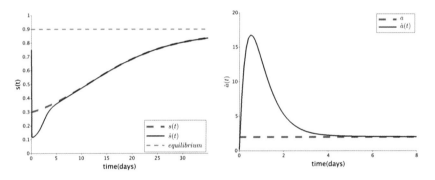

Fig. 5.12 Dynamics of proportion of infected snails $s(t)$ with its estimate $\hat{s}(t)$ (left) and of the estimation $\hat{a}(t)$ of parameter a (right). The green dashed line corresponds to the steady state value of s

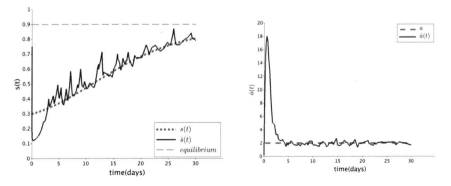

Fig. 5.13 Simulations with measurement noise

Appendix A
Proofs of Some Useful Lemmas

A.1 Proof of Lemma 4.1

The proof is adapted from [8].

Let us first consider pairs $(\bar{A}, \bar{C})$ of the canonical form known as Brunovsky's form

$$\bar{A} = \begin{bmatrix} 0 & \cdots\cdots\cdots & 0 & -a_n \\ 1 & 0 & \cdots\cdots 0 & -a_{n-1} \\ & \ddots & \ddots & \cdots & \cdot \\ \cdot & \cdot & \ddots & \ddots & \cdot \\ \cdot & \cdot & \cdot & 1 & 0 & -a_2 \\ 0 & \cdots\cdots & 0 & 1 & -a_1 \end{bmatrix} , \quad \bar{C} = \begin{bmatrix} 0 & \cdots & 0 & 1 \end{bmatrix}$$

where the a_i are any numbers. Their observability matrices are transposed lower triangular:

$$\bar{O} = \begin{bmatrix} & & 1 \\ 0 & \cdot\cdot \\ & \cdot\cdot & \bigstar \\ 1 & & \end{bmatrix}$$

therefore invertible. It is easy to see that the characteristic polynomial of the matrix $\bar{A}$ is given by

$$\pi_{\bar{A}}(\xi) = \xi^n + a_1 \xi^{n-1} + \cdots + a_{n-1}\xi + a_n.$$

Indeed, if X is a left eigenvector of $\bar{A}$ for an eigenvalue λ (possibly complex), $X\bar{A} = \lambda X$ gives

$$X_2 = \lambda X_1,$$
$$X_3 = \lambda X_2 = \lambda^2 X_1,$$
$$\vdots$$
$$X_n = \lambda X_{n-1} = \lambda^{n-1} X_1,$$
$$-a_n X_1 - a_{n-1} X_2 - \cdots - a_1 X_n = \lambda X_n.$$

Thus the line vector X is of the form

$$X = \begin{bmatrix} 1 & \lambda & \lambda^2 & \cdots & \lambda^{n-1} \end{bmatrix} X_1 \quad \text{with } X_1 \neq 0$$

and λ verifies

$$\left(\lambda^n + a_1 \lambda^{n-1} + a_2 \lambda^{n-2} + \cdots + a_{n-1} \lambda + a_n \right) X_1 = 0 .$$

Since X_1 is non-zero, we deduce that the eigenvalues are roots of the polynomial

$$\lambda^n + a_1 \lambda^{n-1} + a_2 \lambda^{n-2} + \cdots + a_{n-1} \lambda + a_n = 0$$

which is of degree n and whose coefficient of λ^n is equal to 1.

The characteristic polynomial of the matrix $\bar{A} + \bar{G}\bar{C}$, where $\bar{G}$ is a vector of $\mathbb{R}^n$ with elements denoted $\bar{g}_i$, is written as follows

$$\pi_{\bar{A}+\bar{G}\bar{C}}(\xi) = \xi^n + (a_1 - \bar{g}_n)\xi^{n-1} + \cdots + (a_{n-1} - \bar{g}_2)\xi + (a_n - \bar{g}_1).$$

Thus, one can arbitrarily choose the n coefficients of this polynomial by choosing the n elements of $\bar{G}$, and thus freely assign the spectrum of the matrix $\bar{A} + \bar{G}\bar{C}$. For any set $\Lambda = \{\lambda_1, \cdots, \lambda_n\}$ of n real or complex numbers two by two conjugates, one has just to identify the coefficients of the polynomial $\pi_{\bar{A}+\bar{G}\bar{C}}$ with those of

$$\prod_{i=1}^{n} (\xi - \lambda_i) = \xi^n + \sum_{k=1}^{n} (-1)^k \sigma_k(\Lambda)\xi^{n-k}.$$

Thus, we obtain

$$\bar{g}_i = a_{n+1-i} + (-1)^{n-i} \sigma_{n+1-i}(\Lambda), \quad i = 1 \cdots n.$$

Let us now show that for any pair (A, C) such that O is full rank, there is an invertible P matrix such that $P^{-1}AP = \bar{A}$ and $CP = \bar{C}$, where the pair $(\bar{A}, \bar{C})$ is in the Brunovsky's form. Consider the vector

$$L = O^{-1} \begin{bmatrix} 0 \\ \vdots \\ \vdots \\ 1 \end{bmatrix} \Rightarrow \begin{cases} CA^k L = 0, & k = 0 \cdots n - 2, \\ CA^{n-1}L = 1, \end{cases}$$

and the matrix consisting of the concatenation of the columns

$$P = [L \ AL \ \cdots \ A^{n-1}L].$$

We have

$$OL = \begin{bmatrix} 0 \\ \vdots \\ \vdots \\ \vdots \\ 1 \end{bmatrix}, \quad OAL = \begin{bmatrix} CAL \\ CA^2L \\ \vdots \\ CA^{n-1}L \\ CA^n L \end{bmatrix} = \begin{bmatrix} 0 \\ \vdots \\ \vdots \\ 1 \\ \star \end{bmatrix}, \quad OA^2L = \begin{bmatrix} 0 \\ \vdots \\ 1 \\ \star \\ \star \end{bmatrix}, \cdots$$

up to

$$OA^{n-1}L = \begin{bmatrix} 1 \\ \star \\ \vdots \end{bmatrix}.$$

Thus the OP matrix is of the form

$$OP = \begin{bmatrix} 0 & & \iddots & 1 \\ & \iddots & & \\ 1 & \iddots & & \bigstar \end{bmatrix}$$

which shows that P is indeed an invertible matrix. Finally, the columns of the AP matrix are

$$AP = [AL \ A^2L \ \cdots \ A^{n-1}L \ A^n L].$$

Its first $n - 1$ columns are written as follows

$$[AL \ A^2 L \ \cdots \ A^{n-1} L] = P \begin{bmatrix} 0 \cdots & & & \\ 1 & 0 & \cdots & \\ 0 & \ddots & \ddots & \\ \vdots & & 1 & 0 \\ 0 & \cdots & 0 & 1 \end{bmatrix}$$

By Cayley-Hamilton's Theorem, we have $\pi_A(A) = 0$, which allows us to write the last column of AP as

$$A^n L = -a_n L - a_{n-1} AL - \cdots - a_1 A^{n-1} L = P \begin{bmatrix} -a_n \\ -a_{n-1} \\ \vdots \\ -a_1 \end{bmatrix}$$

which shows that we have $PA = \bar{A}P$. Finally, one gets

$$CP = [CL \ CAL \ \cdots \ CA^{n-1} L] = [0 \ \cdots \ 0 \ 1] = \bar{C}.$$

Then, for a vector $\bar{G}$ such that $Sp(\bar{A} + \bar{G}\bar{C}) = \Lambda$, we have $Sp(P^{-1}(\bar{A} + \bar{G}\bar{C})P) = \Lambda$, where $P^{-1}(\bar{A} + \bar{G}\bar{C})P = A + P\bar{G}C$. We conclude that for the gain vector $G = P\bar{G}$, we have $Sp(A + GC) = \Lambda$.

A.2   Proof of Lemma 4.2

The proof is adapted from [35].

Let X be a left eigenvector of $A + GC$ for the eigenvalue λ_i. By writing $X(A + GC) = \lambda_i X$, we obtain the $n - 1$ equalities.

$$\begin{aligned} X_1 &= \lambda_i X_2, \\ X_2 &= \lambda_i X_3, \\ &\vdots \\ X_{n-1} &= \lambda_i X_n. \end{aligned}$$

Thus X_n is necessarily non-zero and can be taken equal to 1, which gives

$$X = \begin{bmatrix} \lambda_i^{n-1} & \lambda_i^{n-2} & \cdots & \lambda_i & 1 \end{bmatrix}.$$

We then obtain the n rows of the matrix $V_{\lambda_1,\cdots,\lambda_n}$, which defines a matrix of change of basis that diagonalizes the matrix $A + GC$.

Now let's show how to determine the inverse of $V_{\lambda_1,\cdots,\lambda_n}$. Let w_{ij} be the coefficients of $V_{\lambda_1,\cdots,\lambda_n}^{-1}$. The equality $(V_{\lambda_1,\cdots,\lambda_n}^{-1})(V_{\lambda_1,\cdots,\lambda_n}) = Id$ gives

$$\sum_{k=1}^{n} w_{kj} \lambda_i^{nk} = \delta_{ij} := \begin{vmatrix} 0 \text{ if } i \neq j, \\ 1 \text{ if } i = j. \end{vmatrix} \tag{A.1}$$

For each j in $\{1, \cdots, n\}$, let's consider the polynomial

$$P_j(X) = \sum_{k=1}^{n} w_{kj} X^{n-k}. \tag{A.2}$$

The conditions (A.1) amount to write $P_j(\lambda_i) = \delta_{ij}$, i.e. the polynomial P_j has $n-1$ roots λ_i for $i \neq j$ and $P_j(\lambda_j)$ is equal to 1. So it has the following expression

$$P_j(X) = \prod_{k \neq j} \frac{X - \lambda_k}{\lambda_j - \lambda_k}.$$

By identifying its coefficients with those of the expression (A.2), we obtain

$$w_{ij} = (-1)^{i-1} \frac{\sigma_{i-1}(\Lambda \setminus \{\lambda_j\})}{\prod_{k \neq j} \lambda_j - \lambda_k} \tag{A.3}$$

where the σ_k are the symmetric functions defined in (4.6).

Let

$$\varphi(\lambda_1, \cdots, \lambda_n) = \lambda_1 + c \|V_{\lambda_1,\cdots,\lambda_n}^{-1}\|_\infty + \theta.$$

The expression (A.3) shows that the norm $\|V_{\lambda_1,\cdots,\lambda_n}^{-1}\|_\infty$ becomes arbitrarily large when $\lambda_i - \lambda_j$ approaches 0 (for $i \neq j$), which ensures the existence of numbers $\lambda_n < \lambda_{n-1} < \cdots < \lambda_1 < 0$ such that $\varphi(\lambda_1, \cdots, \lambda_n) > 0$. For $\lambda_i = -\alpha^i$ ($i = 1, \cdots, n$), we obtain, for any j

$$\lim_{\alpha \to +\infty} w_{ij} = \begin{vmatrix} 0 \ i < n \\ 1 \ i = n \end{vmatrix}$$

and $\|V_{-\alpha,-\alpha^2,\cdots,-\alpha^n}^{-1}\|_\infty$ thus tends towards 1 when α tends towards $+\infty$, which shows the existence of numbers $\lambda_n < \lambda_{n-1} < \cdots < \lambda_1 < 0$ such that $\varphi(\lambda_1, \cdots, \lambda_n) < 0$. Finally, by continuity of φ, we deduce the existence of $\lambda_n < \lambda_{n-1} < \cdots < \lambda_1 < 0$ such that $\varphi(\lambda_1, \cdots, \lambda_n) = 0$.

A.3   Proof of Theorem 4.1

The proof is adapted from [70].

Let P be a symmetric positive definite matrix satisfying $M^\top P + PM + Q = 0$ where Q is a symmetric positive definite matrix. We consider the Lyapunov function

$$V(x) = ||x||_P = x^\top P x$$

whose time derivative along solutions of $\dot{x} = Mx$ is

$$\frac{d}{dt} V(x(t)) = -x(t)^\top Q x(t) \leq - \underbrace{\frac{\lambda_{min}(Q)}{\lambda_{max}(P)}}_{\beta} x(t)^\top P x(t)$$

(where λ_{min}, λ_{max} denote the smallest and largest real eigenvalues of a symmetric matrix). Then, one has

$$\frac{d}{dt} V(x(t)) \leq -\beta V(x(t))$$

where $\beta > 0$. We deduce that $t \mapsto ||x(t)||_P$ converges exponentially to 0.

Conversely, when M is Hurwitz, one can consider the symmetric matrix

$$P = \lim_{t \to +\infty} \int_0^t e^{\tau M^\top} Q e^{\tau M} d\tau$$

(which exists as a sum of terms $t^k e^{\lambda_i t}$ where $Re \lambda_i < 0$). Then, one has

$$M^\top P + PM = \lim_{t \to +\infty} \int_0^t M^\top e^{\tau M^\top} Q e^{\tau M} + e^{\tau M^\top} Q e^{\tau M} M d\tau$$

$$= \lim_{t \to +\infty} \int_0^t \frac{d}{d\tau} \left(e^{\tau M^\top} Q e^{\tau M} \right) d\tau$$

$$= \lim_{t \to +\infty} e^{t M^\top} Q e^{t M} - Q$$

$$= -Q.$$

Let us show that P is necessarily definite positive. If there exists $u \neq 0$ such that $Pu = 0$, then one has

$$u^T P u = 0 = u^T \left(\int_0^\infty e^{\tau M^\top} Q e^{\tau M} d\tau \right) u = \int_0^\infty ||Q^{1/2} e^{\tau M} u||^2 d\tau$$

where $Q^{1/2}$ is the positive definite matrix such that $Q^{1/2}Q^{1/2} = Q$. Therefore, u has to be null which shows that P is non singular.

Finally, if there exists another symmetric definite positive matrix $\tilde{P}$ which satisfies $M^\top \tilde{P} + \tilde{P}M = -Q$, then one has

$$M^\top(\tilde{P} - P) + (\tilde{P} - P)M = 0.$$

But then

$$e^{tM^\top}\left(M^\top(\tilde{P} - P) + (\tilde{P} - P)M\right)e^{tM} = \frac{d}{dt}\left(e^{tM^\top}(\tilde{P} - P)e^{tM}\right) = 0$$

i.e. $t \mapsto e^{tM^\top}(\tilde{P} - P)e^{tM}$ is a constant function. In particular,

$$e^{0M^\top}(\tilde{P} - P)e^{0M} = \tilde{P} - P = \lim_{t \to +\infty} e^{tM^\top}(\tilde{P} - P)e^{tM} = 0$$

which gives $\tilde{P} = P$.

Appendix B
Implementation of the "Boarding School" Example

B.1 Derivation of the Fisher's Information Matrix

In this example, we consider $x \in \mathbb{R}^n$ with $n = 2$, $y \in \mathbb{R}^m$ with $m = 1$, i.e.,

$$x(t) = \begin{bmatrix} S(t) \\ I(t) \end{bmatrix}, \quad y(t) = I(t),$$

and $\theta \in \mathbb{R}^p$ with $p = 2$, i.e.,

$$\theta = \begin{bmatrix} \beta \\ \gamma \end{bmatrix}.$$

We consider the following model, equivalent to model (3.4) with $k = 1$:

$$\begin{cases} \dot{x} = f(x, \theta) = \begin{bmatrix} \dot{S} \\ \dot{I} \end{bmatrix} = \begin{bmatrix} f_S(S, I, \theta) \\ f_I(S, I, \theta) \end{bmatrix} = \begin{bmatrix} -\beta S I/N \\ \beta S I/N - \gamma I \end{bmatrix}, \quad x(0) = x_0 = \begin{bmatrix} S_0 \\ I_0 \end{bmatrix}, \\ y = h(x, \theta) = I. \end{cases}$$

We disregard $\Theta = (\theta, x_0)$ since the initial conditions are assumed to be known in this example. The Jacobian of the observation with respect to the parameter θ is:

$$\chi(t, \theta) = \frac{\partial y}{\partial \theta}(t) = \frac{\partial h}{\partial x} \frac{\partial x}{\partial \theta}(t),$$

since in this example,

$$\frac{\partial h}{\partial \theta}(t) = 0 \, .$$

This Jacobian has dimension $m \times p = 1 \times 2$. We have

$$\frac{\partial h}{\partial x} = \left[\begin{array}{cc} \frac{\partial h}{\partial S} & \frac{\partial h}{\partial I} \end{array} \right] = \left[0 \ 1 \right]$$

and

$$z = \frac{\partial x}{\partial \theta} = \left[\begin{array}{cc} \frac{\partial S}{\partial \beta} & \frac{\partial S}{\partial \gamma} \\ \frac{\partial I}{\partial \beta} & \frac{\partial I}{\partial \gamma} \\ \frac{\partial I}{\partial \beta} & \frac{\partial I}{\partial \gamma} \end{array} \right] \, .$$

This yields

$$\chi = \frac{\partial h}{\partial x} \frac{\partial x}{\partial \theta} = \left[\begin{array}{cc} \frac{\partial I}{\partial \beta} & \frac{\partial I}{\partial \gamma} \end{array} \right] \, .$$

Let $\{t_i\}$, $i = 0, 1, 2, \ldots, M$, be the sampling times. Fisher's Information Matrix is defined as:

$$\mathrm{FIM}(\theta, \sigma) = \frac{1}{\sigma^2} \sum_{i=1}^{M} \chi(t_i, \theta)^{\top} \chi(t_i, \theta) \, ,$$

where σ^2 is defined as the sum of the squared error (SSE) divided with $M - p$ instead of $M - (n + p)$ as in Eq. (5.4), since the initial conditions are assumed to be known in this example.

Let us now compute the Fisher's Information Matrix.
Let

$$A(t) = \frac{\partial f}{\partial x} = \left[\begin{array}{cc} \frac{\partial f_S}{\partial S} & \frac{\partial f_S}{\partial I} \\ \frac{\partial f_I}{\partial S} & \frac{\partial f_I}{\partial I} \end{array} \right] = \left[\begin{array}{cc} -\frac{\beta I}{N} & -\frac{\beta S}{N} \\ \frac{\beta I}{N} & \frac{\beta S}{N} - \gamma \end{array} \right]$$

and

$$B(t) = \frac{\partial f}{\partial \theta} = \left[\begin{array}{cc} \frac{\partial f_S}{\partial \beta} & \frac{\partial f_S}{\partial \gamma} \\ \frac{\partial f_I}{\partial \beta} & \frac{\partial f_I}{\partial \gamma} \end{array} \right] = \left[\begin{array}{cc} -\frac{SI}{N} & 0 \\ \frac{SI}{N} & -I \end{array} \right] \, .$$

The matrix z can be computed by numerically solving the following system of ODE's:

$$\begin{cases} \dot{x} = f(x, \theta), & x(0) = x_0, \\ \dot{z} = Az + B, & z(0) = 0_{n \times p} \end{cases}$$

which is a subsystem of (5.7) since the initial conditions are assumed to be known in this example (i.e., we disregard w). In the following code, the entries of x and z are indexed in this way:

$$x = \begin{bmatrix} x_1 \\ x_2 \end{bmatrix}, \quad z = \begin{bmatrix} z_3 & z_5 \\ z_4 & z_6 \end{bmatrix},$$

leading to

$$\chi = \begin{bmatrix} z_4 & z_6 \end{bmatrix}.$$

B.2   Numerical Implementation

The code has been written with the `Scilab` language and executed under SCILAB `6.0.0`.[1] It consists in a function for identifying β and γ and using the `lsqrsolve` function which implements the Levenberg-Marquardt algorithm to perform ordinary least squares. We could have chosen the `fminsearch` function which is an implementation of the Nelder-Mead algorithm, but this gives exactly the same results. For solving ODE's, `Scilab` uses the `lsoda` solver of ODEPACK. It automatically selects between non-stiff predictor-corrector Adams method and stiff Backward Differentiation Formula (BDF) method. It uses non-stiff method initially and dynamically monitors data in order to decide which method to use.
We define the following functions in the `Scilab` environment:

```
function [kguess_n,SSE]=identifKMK(OBS,T,kguess,N)
// kguess_n = [BETA;GAMMA]
t0=T(1);m=length(T);
[x,SSE]=lsqrsolve(kguess,errorKmcK,m,[1.d-8,1.d-8,
1.d-5,1d9,0,100]);
nbparam=length(kguess)
kguess_n=x;
x0=[N-OBS(1);OBS(1)];
sol=ode(x0,t0,T,list(KmcK,x(1),x(2)));
kguess_n=kguess_n(:);
SSE=sum(SSE.^2)
xset("window",1)
sol1=ode(x0,t0,T(1):0.01:T($),list(KmcK,x(1),x(2)));
```

---

[1] https://www.scilab.org/.

```
clf
plot((T(1):0.01:T($))',sol1(2,:)')
plot(T',OBS,'ro')
endfunction

function y=errorKmcK(k,m)
x0=[N-OBS(1);OBS(1)];x0=x0(:);
BETA=k(1);GAMMA=k(2);
sol=ode(x0,t0,T,list(KmcK,BETA,GAMMA));
predic=sol(2,:);
predic=predic(:);
OBS=OBS(:);
y=OBS-predic;
endfunction

function xdot=KmcK(t,x,BETA,GAMMA,N)
xdot=[-BETA/N*x(2),0;BETA/N*x(2),-GAMMA]*x
endfunction
```

Then, the Scilab session goes like this

```
[commandchars=\\\{\}]
--> load('databoarding')
ans  =
T
--> OBS=dataBSFlu

OBS  =

column 1 to 8

1.   6.   26.   73.   222.   293.   258.   237.

column 9 to 14

191.   124.   68.   26.   10.   3.

--> M=length(OBS);

--> T=0:M-1;

--> N=763;

--> beta0=2;gamma0=0.5;param=[beta0,gamma0];

--> p=length(param);

--> [param,SSE]=identifKMK(OBS,T,param,N)

SSE  =

4892.6472
```

```
param  =

1.9605032
0.4751562
```

```
--> sigma2=SSE/(M-p)

sigma2  =

407.72060
```

```
--> BETA=param(1);GAMMA=param(2);
```

We then compute confidence intervals using the formulas (5.2)–(5.6).

```
[commandchars=\\\{\}]
function FIM=fimKmcK(x0,T,BETA,GAMMA,sigma2)
// Compute the sensitivity matrix
x0=x0(:);t0=T(1);
X0=[x0;0;0;0;0]
sol=ode(X0,t0,T,list(JKmcK,BETA,GAMMA));
M=sol([4,6],:);
FIM=M*M'./sigma2;
endfunction

function xdot=JKmcK(t,x,BETA,GAMMA)
xdot(1)=-BETA*x(1)*x(2)/N
xdot(2)=BETA*x(1)*x(2)/N-GAMMA*x(2)
xdot(3)=-BETA*x(2)*x(3)/N-BETA*x(1)*x(4)/N-x(1)*x(2)/N
xdot(4)=BETA*x(2)*x(3)/N+(BETA*x(1)/N-GAMMA)*x(4)+x(1)
*x(2)/N
xdot(5)=-BETA*x(2)*x(5)/N-BETA*x(1)*x(6)/N
xdot(6)=BETA*x(2)*x(5)/N+(BETA*x(1)/N-GAMMA)*x(6)-x(2)
endfunction
```

Then the Scilab session is

```
[commandchars=\\\{\}]

--> x0=[N-OBS(1); OBS(1)];

--> FIM=fimKmcK(x0,T,BETA,GAMMA)
FIM  =

974.5073    -523.73985
-523.73985    3132.2047
```

```
--> cond(FIM)
ans  =

3.8082403
```

```
--> CovMAT=inv(FIM)
CovMAT  =
```

```
0.0011275    0.0001885
0.0001885    0.0003508

--> t=cdft("T",M-p,0.975,0.025)
t   =

2.1788128

--> confBETA=t*sqrt(CovMAT(1,1))
confBETA   =

0.0731602

--> confGAMMA=t*sqrt(CovMAT(2,2))
confGAMMA   =

0.0408077
```

Appendix C
Implementation of the "Plague in Bombay" Example

C.1 Derivation of the Fisher Information Matrix

In this example, we consider $x \in \mathbb{R}^n$ with $n = 2$, $y \in \mathbb{R}^m$ with $m = 1$, i.e.,

$$x(t) = \begin{bmatrix} S(t) \\ I(t) \end{bmatrix}, \quad y(t) = I(t),$$

and $\theta \in \mathbb{R}^p$ with $p = 2$, i.e.,

$$\tilde{\beta} = \frac{\beta}{N}, \quad \theta = \begin{bmatrix} \tilde{\beta} \\ \gamma \end{bmatrix}.$$

We consider the following model, equivalent to model (3.4) with $k = \gamma$:

$$\begin{cases} \dot{x} = f(x, \theta) = \begin{bmatrix} \dot{S} \\ \dot{I} \end{bmatrix} = \begin{bmatrix} f_S(S, I, \theta) \\ f_I(S, I, \theta) \end{bmatrix} = \begin{bmatrix} -\tilde{\beta} S I \\ \tilde{\beta} S I - \gamma I \end{bmatrix}, \quad x(0) = x_0 = \begin{bmatrix} S_0 \\ I_0 \end{bmatrix}, \\ y = h(x, \theta) = \gamma I. \end{cases}$$

We consider $\Theta = (\theta, x_0)$ since the initial conditions are assumed to be unknown in this example. The Jacobian of the observation with respect to the parameter Θ is:

$$\chi(t, \Theta) = \frac{\partial y}{\partial \Theta}(t) = \frac{\partial h}{\partial x} \frac{\partial x}{\partial \Theta}(t) + \frac{\partial h}{\partial \Theta}(t),$$

© The Author(s), under exclusive license to Springer Nature Singapore Pte Ltd. 2024
N. Cunniffe et al., *Identifiability and Observability in Epidemiological Models*,
SpringerBriefs on PDEs and Data Science,
https://doi.org/10.1007/978-981-97-2539-7

This Jacobian has dimension $m \times (p + n) = 1 \times 4$. We have

$$\frac{\partial h}{\partial x} = \left[\frac{\partial h}{\partial S} \; \frac{\partial h}{\partial I} \right] = \left[0 \; \gamma \right],$$

$$\frac{\partial h}{\partial \Theta} = \left[\frac{\partial h}{\partial \tilde{\beta}} \; \frac{\partial h}{\partial \gamma} \; \frac{\partial h}{\partial S_0} \; \frac{\partial h}{\partial I_0} \right] = \left[0 \; I \; 0 \; 0 \right],$$

and

$$\frac{\partial x}{\partial \Theta} = \begin{bmatrix} \dfrac{\partial S}{\partial \tilde{\beta}} & \dfrac{\partial S}{\partial \gamma} & \dfrac{\partial S}{\partial S_0} & \dfrac{\partial S}{\partial I_0} \\[2mm] \dfrac{\partial I}{\partial \tilde{\beta}} & \dfrac{\partial I}{\partial \gamma} & \dfrac{\partial I}{\partial S_0} & \dfrac{\partial I}{\partial I_0} \\[2mm] \dfrac{\partial \tilde{\beta}}{\partial \tilde{\beta}} & \dfrac{\partial }{\partial \gamma} & \dfrac{\partial }{\partial S_0} & \dfrac{\partial }{\partial I_0} \end{bmatrix}.$$

This yields

$$\frac{\partial h}{\partial x} \frac{\partial x}{\partial \tilde{\theta}} = \gamma \left[\frac{\partial I}{\partial \tilde{\beta}} \; \frac{\partial I}{\partial \gamma} \; \frac{\partial I}{\partial S_0} \; \frac{\partial I}{\partial I_0} \right].$$

Therefore,

$$\chi = \gamma \left[\frac{\partial I}{\partial \tilde{\beta}} \; \frac{\partial I}{\partial \gamma} + I \; \frac{\partial I}{\partial S_0} \; \frac{\partial I}{\partial I_0} \right].$$

Let $\{t_i\}$, $i = 0, 1, 2, \ldots, M$, be the sampling times. Fisher's Information Matrix is defined as:

$$\mathrm{FIM}(\Theta, \sigma) = \frac{1}{\sigma^2} \sum_{i=1}^{M} \chi(t_i, \Theta)^{\top} \chi(t_i, \Theta),$$

where σ^2 is defined as in Eq. (5.4), since the initial conditions are unknown in this example.

We compute now the Fisher's Information Matrix. We make the following decomposition:

$$\frac{\partial x}{\partial \Theta} = \left[\frac{\partial x}{\partial \theta} \; \frac{\partial x}{\partial x_0} \right] = \left[z \; w \right],$$

with

$$z = \frac{\partial x}{\partial \theta} = \begin{bmatrix} \dfrac{\partial S}{\partial \tilde{\beta}} & \dfrac{\partial S}{\partial \gamma} \\[2mm] \dfrac{\partial I}{\partial \tilde{\beta}} & \dfrac{\partial I}{\partial \gamma} \\[2mm] \dfrac{\partial }{\partial \tilde{\beta}} & \dfrac{\partial }{\partial \gamma} \end{bmatrix} \quad \text{and} \quad w = \frac{\partial x}{\partial x_0} = \begin{bmatrix} \dfrac{\partial S}{\partial S_0} & \dfrac{\partial S}{\partial I_0} \\[2mm] \dfrac{\partial I}{\partial S_0} & \dfrac{\partial I}{\partial I_0} \\[2mm] \dfrac{\partial }{\partial S_0} & \dfrac{\partial }{\partial I_0} \end{bmatrix}.$$

Letting

$$A(t) = \frac{\partial f}{\partial x} = \begin{bmatrix} \dfrac{\partial f_S}{\partial S} & \dfrac{\partial f_S}{\partial I} \\[2mm] \dfrac{\partial f_I}{\partial S} & \dfrac{\partial f_I}{\partial I} \end{bmatrix} = \begin{bmatrix} -\tilde{\beta}I & -\tilde{\beta}S \\ \tilde{\beta}I & \tilde{\beta}S - \gamma \end{bmatrix},$$

and

$$B(t) = \frac{\partial f}{\partial \theta} = \begin{bmatrix} \dfrac{\partial f_S}{\partial \tilde{\beta}} & \dfrac{\partial f_S}{\partial \gamma} \\[2mm] \dfrac{\partial f_I}{\partial \tilde{\beta}} & \dfrac{\partial f_I}{\partial \gamma} \end{bmatrix} = \begin{bmatrix} -SI & 0 \\ SI & -I \end{bmatrix}.$$

The FIM can be computed via numerically solving the following system of ODE's:

$$\begin{cases} \dot{x} = f(x, \theta), & x(0) = x_0, \\ \dot{z} = Az + B, & z(0) = 0_{n \times p}, \\ \dot{w} = Aw, & w(0) = \mathrm{Id}_{n \times n}, \end{cases}$$

which repeats Eq. (5.7). In the following code, the entries of x, z, and w are indexed in this way:

$$x = \begin{bmatrix} x_1 \\ x_2 \end{bmatrix}, \quad z = \begin{pmatrix} z_3 & z_5 \\ z_4 & z_6 \end{pmatrix}, \quad w = \begin{bmatrix} w_7 & w_9 \\ w_8 & w_{10} \end{bmatrix},$$

leading to

$$\chi = \gamma \begin{bmatrix} z_4 & z_6 + x_2 & w_8 & w_{10} \end{bmatrix}.$$

C.2   Numerical Implementation

Although the code is very similar the one provided in the previous example (Appendix B), we provide it for convenience, as it required a number of small changes.
We define the following functions in the Scilab environment:

```
function [kguess_n ,SSE]=identifKMK(OBS,T,kguess)
t0=T(1);m=length(T);
[x,SSE]=lsqrsolve(kguess,errorKmcK,m,[1.d-8,1.d-8,
1.d-5,1d9,0,100]);
nbparam=length(kguess);
kguess_n=x;
x0=[kguess_n(3);kguess_n(4)];
sol=ode(x0,t0,T,list(KmcK,x(1),x(2)));
kguess_n=kguess_n(:);
SSE=sum(SSE.^2);
xset("window",1);
```

```
sol1=ode(x0,t0,T(1):0.01:T($),list(KmcK ,x(1),x(2)));
clf;
plot((T(1):0.01:T($))',kguess_n(2)*sol1(2,:)','k');
plot(T',OBS ,'ko')
legend(["$\Large \gamma I(t)$", "$\Large\mbox{Data}$"])
ylabel("$\Large \mbox{number of deaths per week}$",
"fontsize",3);
xlabel("$\Large \mbox{time }t\mbox{ (in week)}$",
"fontsize",3);
endfunction

function y=errorKmcK(k,m)
x0=[k(3);k(4)];
x0=x0(:);
B=k(1);
GAMMA=k(2);
sol=ode(x0,t0,T,list(KmcK,B,GAMMA));
predic=GAMMA*sol(2,:);
predic=predic(:);
OBS=OBS(:);
y=OBS-predic;
endfunction

function  xdot=KmcK(t,x,B,GAMMA)
xdot=[-B*x(2),0; B*x(2),-GAMMA]*x
endfunction
```

Then, the Scilab session goes like this

```
[commandchars=\\\{\}]
--> load('databombay')
ans  =
T
--> OBS=dataBSFlu

OBS  =

column 1 to 9

8.    10.    12.    16.    24.    48.    51.    92.    124.

column 10 to 16

178.    280.    387.    442.    644.    779.    702.

column 17 to 23

695.    870.    925.    802.    578.    404.    296.

column 24 to 31

162.    106.    64.    46.    35.    27.    28.    24.
```

```
--> M=length(OBS);

--> T=0:M-1;

--> S0=15000;I0=7;gamma0=0.6;b0=8e-5;//Initial guesses

--> param=[b0,gamma0,S0,I0];//With b=beta/N

--> p=length(param);

--> [param,SSE]=identifKMK(OBS,T,param,N)

SSE  =

106336.49

param  =

0.0000855
3.7161743
48113.13
1.4213612

--> n=2;sigma2=SSE/(M-(n+p))

sigma2  =

4253.4597

--> B=param(1);GAMMA=param(2);
```

We then compute confidence intervals using the formulas (5.2)–(5.6).

```
[commandchars=\\\{\}]
function  FIM=fimKmcK(x0,T,B,GAMMA,sigma2)
// Compute  the  sensitivity  matrix
x0=x0(:);t0=T(1);
X0=[x0;zeros(4,1);eye(2,2)(:)];
sol=ode(X0,t0,T,list(JKmcK,B,GAMMA));
M=GAMMA*sol([4,6,8,10],:);
M(2,:)=M(2,:)+GAMMA*sol(2,:);
FIM=M*M'./sigma2;
endfunction

function  xdot=JKmcK(t,x,B,GAMMA)
xdot (1)=-B*x(1)*x(2);
xdot (2)=B*x(1)*x(2)-GAMMA*x(2);

xdot (3)=-B*x(2)*x(3)-B*x(1)*x(4)-x(1)*x(2);
xdot (4)=B*x(2)*x(3)+(B*x(1)-GAMMA)*x(4)+x(1)*x(2);
xdot (5)=-B*x(2)*x(5)-B*x(1)*x(6);
xdot (6)=B*x(2)*x(5)+(B*x(1)-GAMMA)*x(6)-x(2);

xdot (7)=-B*x(2)*x(7)-B*x(1)*x(8);
xdot (8)=B*x(2)*x(7)+(B*x(1)-GAMMA)*x(8);
xdot (9)=-B*x(2)*x(9)-B*x(1)*x(10);
```

```
xdot (10)=B*x(2)*x(9)+(B*x(1)-GAMMA)*x(10);
endfunction
```

Then the Scilab session is

```
[commandchars=\\\{\}]

--> x0=[param(3);param(4)];

--> FIM=fimKmcK(x0,T,B,GAMMA)
FIM  =

1.100D+14   -2.117D+09    196924.7    92108921.
-2.117D+09   40885.251   -3.7835134  -1826.2845
196924.7    -3.7835134   0.0003533   0.1603199
92108921.   -1826.2845   0.1603199   118.22324

--> cond(FIM)
ans  =

9.141D+24

--> CovMAT=inv(FIM)
Warning: Matrix is close to singular or badly scaled.
CovMAT  =

0.0000006    0.0093978   -220.96547   -0.0128142
0.0093978    150.37071   -3535242.5   -204.997
-220.96547   -3535242.5   8.313D+10    4820745.
-0.0128142   -204.997     4820745.     279.63056

--> t=cdft("T",M-(n+p),0.975,0.025)
t  =

2.0595386

--> confB=t*sqrt(CovMAT(1,1))
confB  =

0.0015784

--> confGAMMA=t*sqrt(CovMAT(2,2))
confGAMMA  =

25.255243

--> confS0=t*sqrt(CovMAT(3,3))
confS0  =

593794.26

--> confI0=t*sqrt(CovMAT(4,4))
confI0  =

34.439929
```

Appendix D
Generalized Least Squares

With Ordinary Least Squares, constant variance has been assumed which may be not appropriate for some data. A relative error, i.e., when the error is assumed to be proportional to the size of the measurement, is an assumption that might be reasonable when counting individuals in a population.
In this case we assume that the observations are [14, 16, 28, 37]:

$$Y_i = y(t_i, \Theta) + y(t_i, \Theta)^\rho \, \mathcal{E}_i \, ,$$

with $\Theta = (\theta, x_0)$ and ρ will be made precise later.
The criterion to be minimized is

$$\mathcal{J}(\Theta) = \sum_{i=1}^{N} w_i \, [Y_i - y(t_i, \Theta)]^2 \, .$$

The values of the weights (w_i) depend on the value of the model and are not known. The process is carried with an iterated re-weighted least squares:

1. Estimate $\hat{\Theta}_0$ with an OLS step ($\rho = 0$): $w_i = 1$ for all $i = 1, \ldots, N$
2. Set $\rho = 1$ (for instance) and $w_i = 1 / \left[y(t_i, \hat{\Theta}_0) \right]^{2\rho}$ for all $i = 1, \ldots, N$
3. Form $\mathcal{J}(\Theta)$ with these w_i and estimate

$$\hat{\Theta}_1 = \arg\min_{\Theta} \mathcal{J}(\Theta)$$

4. Continue the procedure till the estimates $\hat{\Theta}_k$ and $\hat{\Theta}_{k+1}$ are sufficiently close to each other, to obtain $\hat{\Theta}_{\text{GLS}}$.

© The Author(s), under exclusive license to Springer Nature Singapore Pte Ltd. 2024   101
N. Cunniffe et al., *Identifiability and Observability in Epidemiological Models*,
SpringerBriefs on PDEs and Data Science,
https://doi.org/10.1007/978-981-97-2539-7

With $\hat{\Theta}_{\text{GLS}}$, as in the OLS case, we can obtain the covariance matrix (Σ) of the estimated parameters, approximated as the inverse Fisher Information Matrix with weights:

$$\text{FIM}(\hat{\Theta}_{\text{GLS}}, \hat{\sigma}^2_{\text{GLS}}) = \frac{1}{\hat{\sigma}^2_{\text{GLS}}} \sum_{i=1}^{N} \frac{1}{y(t_i, \hat{\Theta}_{\text{GLS}})^{2\rho}} \frac{\partial y}{\partial \Theta}(t_i, \hat{\Theta}_{\text{GLS}})^{\top} \frac{\partial y}{\partial \Theta}(t_i, \hat{\Theta}_{GLS}),$$

with

$$\hat{\sigma}^2_{\text{GLS}} = \frac{1}{N - p} \sum_{i=1}^{N} \frac{1}{y(t_i, \hat{\Theta}_{\text{GLS}})^{2\rho}} \left[Y_i - y(t_i, \hat{\Theta}_{\text{GLS}}) \right]^2.$$

We then obtain

$$\hat{\Sigma}_{\text{GLS}} = \left[\text{FIM}(\hat{\Theta}_{\text{GLS}}, \hat{\sigma}^2_{\text{GLS}}) \right]^{-1},$$

The square roots of the diagonal elements of the approximation of the covariance matrix $\hat{\Sigma}_{\text{GLS}}$ give the standard errors.

References

1. Abdelhedi, A., Boutat, D., Sbita, L., Tami, R., Liu, D.-Y.: Observer design for a class of nonlinear piecewise systems. Application to an epidemic model with treatment. Math. Biosci. **271**, 128–135 (2016)
2. Aboky, C., Sallet, G., Vivalda, J.C.: Observers for Lipschitz nonlinear systems. Int. J. Contr. **75**(3), 204–212 (2002)
3. Aeyels, D.: Generic observability of differentiable systems. SIAM J. Control Optim. **19**, 595–603 (1981)
4. Aeyels, D.: On the number of samples necessary to achieve observability. Syst. Control Lett. **1**, 92–94 (1981–1982)
5. Akaike, H.: A new look at the statistical model identification. IEEE Trans. Autom. Control **19**, 716–723 (1976)
6. Alonso-Quesada, S., De la Sen, M., Agarwal, R.P., Ibeas, A.: An observer-based vaccination control law for an SEIR epidemic model based on feedback linearization techniques for nonlinear systems. Adv. Differ. Equ. **2012**, 161 (2012)
7. Anderson, R.M., May, R.M.: Infectious Diseases of Humans. Dynamics and Control. Oxford Science Publications, Oxford (1991)
8. Andréa-Novell, B., de Lara, M.: Control Theory for Engineers: A Primer. Springer, Berlin (2013)
9. Andrieu, V., Besançon, G., Serres, U.: Observability necessary conditions for the existence of observers. In: IEEE Conference on Decision and Control, Florence (2013)
10. Andrieu, V., Praly, L.: On the existence of a Kazantzis–Kravaris/Luenberger observer. SIAM J. Control Optim. **45**(2), 432–456 (2006)
11. Antsaklis, P.J., Michel, A.N.: State Feedback and State Observers, pp. 351–410. Birkhäuser, Boston (2007)
12. Bacaer, N.: The model of Kermack and McKendrick for the plague epidemic in Bombay and the type reproduction number with seasonality. J. Math. Biol. **64**, 403–422 (2012)
13. Banks, H.T., Cintrón-Arias, A., Kappel, F.: Parameter selection methods in inverse problem formulation. In: Mathematical Modeling and Validation in Physiology. Lecture Notes in Mathematics, vol. 2064, pp. 43–73. Springer, Heidelberg (2013)
14. Banks, H.T., Davidian, M., Samuels, J., Sutton, K.: An inverse problem statistical methodology summary. In: Chowell, G. (ed.) Mathematical and Statistical Estimation Approaches in Epidemiology, pp. 249–302. Springer, Berlin (2009)

15. Banks, H.T., Ernstberger, S.L., Grove, S.L.: Standard errors and confidence intervals in inverse problems: sensitivity and associated pitfalls. J. Inverse Ill-Posed Probl. **15**, 1–18 (2007)
16. Banks, H.T., Hu, S., Thompson, W.C.: Modeling and Inverse Problems in the Presence of Uncertainty. Monographs and Research Notes in Mathematics (CRC Press, Boca Raton, 2014)
17. Barbour, A.D.: Macdonald's model and the transmission of bilharzia. Trans. R. Soc. Trop. Med. Hyg. **72**, 6–15 (1978)
18. Bellman, R., Åström, K.J.: On structural identifiability. Math. Biosci. **7**, 329–339 (1970)
19. Bellu, G., Saccomani, M.P., Audoly, S., D'Angio, L.: Daisy: a new software tool to test global identifiability of biological and physiological systems.. Comput. Methods Programs Biomed. **88**, 52–61 (2007)
20. Bichara, D., Cozic, N., Iggidr, A.: On the estimation of sequestered infected erythrocytes in plasmodium falciparum malaria patients. Math. Biosci. Eng. **11**, 741–759 (2014)
21. Bichara, D., Guiro, A., Iggidr, A., Ngom, D.: State and parameter estimation for a class of schistosomiasis models. Math. Biosci. **315**, 1–11 (2019)
22. Bichara, D., Iggidr, A., Oumoun, M., Rapaport, A., Sallet, G.: Identifiability and Observability via decoupled variables: application to a malaria intra-host model. In: Proceedings of the 22nd Triennal IFAC World Congress, Yokohama (2023)
23. Bolker, B.M.: Ecological Models and Data in R. Princeton University Press, Princeton (2008)
24. Brauer, F., Castillo-Chávez, C.: Mathematical Models in Population Biology and Epidemiology. Texts in Applied Mathematics, vol. 40. Springer, New York (2001)
25. Brauer, F., Castillo-Chavez, C., Feng, Z.: Mathematical Models in Epidemiology. Texts in Applied Mathematics, vol. 69. Springer, New York (2019). With a foreword by Simon Levin
26. Brauer, F., Wu, J., van den Driessche, P. (eds.): Mathematical Epidemiology. Lectures Notes in Mathematics, no. 1945. Springer, Berlin (2008)
27. Cantó, B.N., Coll, C., Sánchez, E.: Estimation of parameters in a structured SIR model. Adv. Differ. Equ. **2017**, 33 (2017)
28. A. Capaldi, Behrend, S., Berman, B., Smith, J., Wright, J., Lloyd, A.L.: Parameter estimation and uncertainty quantification for an epidemic model. Math. Biosci. Eng. **9**, 553–576 (2012)
29. Capistran, M.A., Moreles, M.A., Lara, B.: Parameter estimation of some epidemic models. The case of recurrent epidemics caused by respiratory syncytial virus. Bull. Math. Biol. **71**, 1890–1901 (2009)
30. Casti, J.L.: Recent developments and future perspectives in nonlinear system theory. SIAM Rev. **24**(3), 301–331 (1982)
31. C. D. S. Center: Influenza in a boarding school. British Med. J. **1**, 587 (1978)
32. Chen, W., Saif, M.: Unknown input observer design for a class of nonlinear systems: an LMI approach. In: *American Control Conference, Minneapolis* (2006)
33. Chiş, O., Banga, J.R., Balsa-Canto, E.: Genssi: a software toolbox for structural identifiability analysis of biological models. Bioinformatics **27**, 2610–2611 (2011)
34. Chis, O.T., Villaverde, A.F., Banga, J.R., Balsa-Canto, E.: On the relationship between sloppiness and identifiability. Math. Biosci. **282**, 147–161 (2016)
35. Ciccarella, G., Dala Maara, M., Germani, A.: A Luenberger-like observer for nonlinear systems. Int. J. Control **57**, 537–556 (1993)
36. Cintrón-Arias, A., Banks, H.T., Capaldi, A., Lloyd, A.L.: A sensitivity matrix based methodology for inverse problem formulation. J. Inverse Ill-Posed Probl. **17**, 545–564 (2009)
37. Cintrón-Arias, A., Castillo-Chávez, C., Bettencourt, L.M.A., Lloyd, A.L., Banks, H.T.: The estimation of the effective reproductive number from disease outbreak data. Math. Biosci. Eng. **6**, 261–282 (2009)
38. Cole, D.J.: Parameter Redundancy and Identifiability. Chapman and Hall/CRC, Boca Raton (2020)
39. Commission: XXII-epidemiological observations made by the commission in bombay city. J. Hyg. **7**, 724–798 (1907)

40. Daley, D., Gani, J.: Epidemic Modelling : An Introduction. Cambridge University Press, Cambridge (1999)
41. Davidian, M., Giltinan, D.: Nonlinear Models for Repeated Measurement Data. Chapman and Hall, Boca Raton (1995)
42. de Vries, G., Hillen, T., Lewis, M., Müller, J., Schönfisch, B.: A Course in Mathematical Biology. Mathematical Modeling and Computation, vol. 12. Society for Industrial and Applied Mathematics (SIAM), Philadelphia (2006). Quantitative modeling with mathematical and computational methods
43. Demmel, J.W.: On condition numbers and the distance to the nearest ill-posed problem. Numer. Math. **51**, 251–289 (1987)
44. Diaby, M., Iggidr, A., Sy, M.: Observer design for a schistosomiasis model. Math. Biosci. **269**, 17–29 (2015)
45. Diop, S., Fliess, M.: Nonlinear observability, identifiability, and persistent trajectories. In: Proceedings 36th IEEE Conference on Decision and Control (CDC), pp. 714–719 (1991)
46. Diop, S., Fliess, M.: On nonlinear observability. In: Proceedings EEC91, vol. 1, pp. 154–211. Hermès, Paris (1991)
47. Diop, S., Wang, Y.: Equivalence between algebraic observability and local generic observability. In: Proceedings of 32nd IEEE Conference on Decision and Control. IEEE, Piscataway (1993)
48. Distefano, J., Cobelli, C.: On parameter and structural identifiability: nonunique observability/reconstructibility for identifiable systems, other ambiguities and new definitions. IEEE Trans. Autom. Control **25**, 830–833 (1980)
49. DiStefano, J.: Dynamic Systems Biology Modeling and Application. Academic Press, Cambridge (2013)
50. Eisenberg, M.C., Robertson, S.L., Tien, J.H.: Identifiability and estimation of multiple transmission pathways in cholera and waterborne disease. J. Theor. Biol. **324**, 84–102 (2013)
51. Evans, N.D., White, L.J., Chapman, M.J., Godfrey, K.R., Chappell, M.J.: The structural identifiability of the susceptible infected recovered model with seasonal forcing. Math. Biosci. **194**, 175–197 (2005)
52. Fliess, M.: Nonlinear control theory and differential algebra. In: Modelling and Adaptive Control (Sopron, 1986). Lecture Notes in Control and Information Sciences, vol. 105, pp. 134–145. Springer, Berlin (1988)
53. Fliess, M., Glad, T.: Essays on Control. Springe Sciences, chap. An Algebraic Approach to Linear and Nonlinear Control, vol. 8, pp. 223–267. Springer, Berlin (1993)
54. Gauthier, J.P., Kupka, I.A.K.: Deterministic Observation Theory and Applications. Cambridge University Press, Cambridge (2001)
55. Golub, G.H., Van Loan, C.: Matrix Computations. The John Hopkins Univesity Press, Baltimore (1989)
56. Gravenor, M.B., McLean, A.R., Kwiatkowski, D.: The regulation of malaria parasitaemia: parameter estimates for a population model. Parasitology **110**(Pt 2), 115–22 (1995)
57. Gravenor, M.B., van Hensbroek, M.B., Kwiatkowski, D.: Estimating sequestered parasite population dynamics in cerebral malaria. Proc. Natl. Acad. Sci. USA **95**, 7620–7624 (1998)
58. Griffith, E.W., Kumar, K.S.P.: On the observability of nonlinear systems. I. J. Math. Anal. Appl. **35**, 135–147 (1971)
59. Hadeler, K.P.: Parameter identification in epidemic models. Math. Biosci. **229**, 185–189 (2011)
60. Hammouri, H., Gauthier, J.P., Othman, S.: A simple observer for nonlinear systems applications to bioreactors. IEEE Trans. Autom. Control **37**, 875–880 (1992)
61. Hartman, P.: Ordinary Differential Equations. Classics in Applied Mathematics, vol. 38. Society for Industrial and Applied Mathematics (SIAM), Philadelphia (2002). Corrected reprint of the second (1982) edition [Birkhäuser, Boston, MA; MR0658490 (83e:34002)], With a foreword by Peter Bates
62. Hermann, R., Krener, A.J.: Nonlinear Controllability and Observability. IEEE Trans. Autom. Control **22**, 728–740 (1977)

63. Hethcote, H.W., Van Ark, J.W.: Epidemiological models for heterogeneous populations: proportionate mixing, parameter estimation, and immunization programs. Math. Biosci. **84**, 85–118 (1987)
64. Higham, N.J.: Matrix nearness problems and applications. In: Applications of Matrix Theory (Bradford, 1988). Institute of Mathematics and its Applications Conference Series, vol. 22, pp. 1–27. Oxford University Press, New York (1989)
65. Hong, H., Ovchinnikov, A., Pogudin, G., Yap, C.: SIAN: a tool for assessing structural identifiability of parametric ODEs. ACM Commun. Comput. Algebra **53**, 37–40 (2019)
66. Hou, M., Müller, P.C.: Design of observers for linear systems with unknown inputs. IEEE Trans. Autom. Control **37**, 871–875 (1992)
67. Inouye, Y.: On the observability of autonomous nonlinear systems. J. Math. Anal. Appl. **60**(1), 236–247 (1977)
68. Jacquez, J.A., Greif, P.: Numerical paremeter identifiability and estimability: integrating identifiability, estimability and optimal sampling design. Math. Biosci. **77**, 201–277 (1985)
69. Kailath, T.: Linear Systems. Prentice-Hall, Upper Saddle River (1990)
70. Khalil, H.: Nonlinear Systems. Prentice-Hall, Upper Saddle River (1996)
71. Kalman, R.E.: Mathematical description of linear dynamical systems. J. SIAM Control Ser. A **1**, 152–192 (1963)
72. Kemper, J.T.: Identification of Silent Infections in SIR Epidemics. Bull. Math. Biol. **43**, 249–257 (1981)
73. Kermack, W., McKendrick, A.: A contribution to the mathematical theory of epidemics. Proc. R. Soc. **A115**, 700–721 (1927)
74. Korobeinikov, A., Wake, G.C.: Lyapunov functions and global stability for SIR, SIRS, and SIS epidemiological models. Appl. Math. Lett. **151**(8), 955–960 (2002)
75. Kazantzis, N., Kravaris, C.: Nonlinear observer design using Lyapunov's auxiliary theorem. Syst. Control Lett. **34**, 241–247 (1998)
76. Le Dimet, F.X., Talagrand, O.: Variational algorithms for analysis and assimilation of meteorological observations: theoretical aspects. Tellus **38A**, 97–110 (1986)
77. Leander, J., Lundh, T., Jirstrand, M.: Stochastic differential equations as a tool to regularize the parameter estimation problem for continuous time dynamical systems given discrete time measurements. Math. Biosci. **251**, 54–62 (2014)
78. Levant, A.: Higher-order sliding modes, differentiation and output-feedback control. Int. J. Autom. **76**(9–10), 924–941 (2003)
79. Li, M., Dushoff, J., Bolker, B.M.: Fitting mechanistic epidemic models to data: a comparison of simple Markov chain Monte Carlo approaches. Stat. Methods Med. Res. **27**, 1956–1967 (2018)
80. Li, M.Y.: An Introduction to Mathematical Modeling of Infectious Diseases. Mathematics of Planet Earth, vol. 2. Springer, Cham (2018)
81. Lintusaari, J., Gutmann, M.U., Kaski, S., Corander, J.: On the identifiability of transmission dynamic models for infectious diseases. Genetics **202**, 911–918 (2016)
82. Ljung, L.: System Identification: Theory for the User. Prentice Hall, Upper Saddle River (1999)
83. Ljung, L., Glad, T.: On global identifiability for arbitrary model parametrizations. Autom. J. **30**, 265–276 (1994)
84. Luenberger, D.G.: An introduction to observers. IEEE Trans. Autom. Control **16**, 596–602 (1971)
85. Magal, P., Webb, G.: The parameter identification problem for SIR epidemic models: identifying unreported cases. J. Math. Biol. **77**, 1629–1648 (2018)
86. Martcheva, M.: An Introduction to Mathematical Epidemiology. Texts in Applied Mathematics, vol. 61. Springer, New York (2015)
87. Miao, H., Xia, X., Perelson, A.S., Wu, H.: On identifiability of nonlinear ODE models and applications in viral dynamics. SIAM Rev. **53**, 3–39 (2011)
88. Moreno, J., Dochain, D.: Global observability and detectability analysis of uncertain reaction systems. Int. J. Control **81**(7), 1062–1070 (2008)

89. Murray, J.: Mathematical Biology I: An introduction. Interdisciplinary Applied Mathematics, vol. 17. Springer, Berlin (2002)
90. Nazari, S.: The unknown input observer and its advantages with examples(2015, preprint). arXiv:1504.07300
91. Nguyen, V.K., Binder, S.C., Boianelli, A., Meyer-Hermann, M., Hernandez-Vargas, E.A.: Ebola virus infection modeling and identifiability problems. Front. Microbiol. **6**, 257 (2015)
92. Nowak, M.A., May, R.M.: Virus Dynamics. Mathematical Principles of Immunology and Virology. Oxford University Press, Oxford (2000)
93. Ollivier, F.: Le problème de l'identifiabilité structurelle globale: étude théorique, méthodes effectives et bornes de complexité. Ph.D. Thesis, Ecole Polytechnique (1990)
94. O'Neill, P.D.: A tutorial introduction to bayesian inference for stochastic epidemic models using Markov chain Monte Carlo methods. Math. Biosci. **180**, 103–114 (2002)
95. Ovchinnikov, A., Pillay, A., Pogudin, G., Scanlon, T.: Multi-experiment parameter identifiability of odes and model theory. SIAM J. on Applied Algebra and Geometry **6**(3), 339–367 (2022)
96. Pant, S.: Information sensitivity functions to assess parameter information gain and identifiability of dynamical systems. J. R. Soc. Interface **15**, 1–20 (2018)
97. Parks, P.C.: A. M. Lyapunov's stability theory–100 years on. IMA J. Math. Control Inf. **9**(4), 275–303 (1992)
98. Peifer, M., Timmer, J.: Parameter estimation in ordinary differential equations for biochemical processes using the method of multiple shooting. IET Syst. Biol. **1**(2), 78–88 (2007)
99. Perasso, A., Laroche, B., Chitour, Y., Touzeau, S.: Identifiability analysis of an epidemiological model in a structured population. J. Math. Anal. Appl. **374**, 154–165 (2011)
100. Perelson, A.S., Nelson, P.W.: Mathematical analysis of HIV-1 dynamics in vivo. SIAM Rev. **41**(1), 3–44 (1999)
101. Raissi, M., Ramezani, N., Seshaiyer, P.: On parameter estimation approaches for predicting disease transmission through optimization, deep learning and statistical inference methods. Lett. Biomath. **6**, 26 (2019)
102. Rapaport, A., Maloum, A.: Design of exponential observers for nonlinear systems by embedding. Int. J. Robust Nonlinear Control **14**, 273–288 (2004)
103. Raue, A., Kreutz, C., Maiwald, T., Bachmann, J., Schilling, M., Klingmuller, U., Timmer, J.: Structural and practical identifiability analysis of partially observed dynamical models by exploiting the profile likelihood. Bioinformatics **25**, 1923–1929 (2009)
104. Reid, J.: Structural identifiability in linear-time invariant systems. IEEE Trans. Autom. Control **22**, 242–246 (1977)
105. Roda, W.C.: Bayesian inference for dynamical systems. Infect. Dis. Model. **5**, 221–232 (2020)
106. Roda, W.C., Varughese, M.B., Han, D., Li, M.Y.: Why is it difficult to accurately predict the covid-19 epidemic? Infect. Dis. Model. **5**, 271–281 (2020)
107. Saccomani, M.P.: An effective automatic procedure for testing parameter identifiability of HIV/AIDS models. Bull. Math. Biol. **73**, 1734–1753 (2011)
108. Saccomani, M.P., Audoly, S., D'Angiò, L.: Parameter identifiability of nonlinear systems: the role of initial conditions. Autom. J. IFAC **39**, 619–632 (2003)
109. Seber, G.A.F., Wild, C.J.: Nonlinear Regression. Wiley Series in Probability and Mathematical Statistics: Probability and Mathematical Statistics. Wiley, New York (1989)
110. Sontag, E.D.: Mathematical Control Theory, Deterministic Finite Dimensional Systems. Texts in Applied Mathematics, vol. 6. Springer, Berlin (1990)
111. Sontag, E.D.: For differential equations with r parameters, $2r + 1$ experiments are enough for identification. J. Nonlinear Sci. **12**, 553–583 (2002)
112. Sontag, E.D., Wang, Y.: I/O equations for nonlinear systems and observation spaces. In: Proceedings Under IEEE Conference on Decision, Brighton, pp. 720–725. IEEE Publications, Piscataway (1991)
113. Spear, R.C., Hubbard, A.: Parameter estimation and site-specific calibration of disease transmission models. Adv. Exp. Med. Biol. **673**, 99–111 (2010)
114. Spurgeon, S.: Sliding mode observers: a survey. Int. J. Syst. Sci. **39**(8), 751–764 (2008)

115. Talagrand, O.: On the mathematics of data assimilation. Tellus **33**, 321–339 (1981)
116. Tarantola, A.: Inverse Problem Theory and Methods for Model Parameter Estimation. Society for Industrial and Applied Mathematics (SIAM), Philadelphia (2005)
117. Tönsing, C., Timmer, J., Kreutz, C.: Profile likelihood-based analyses of infectious disease models. Stat. Methods Med. Res. **27**, 1979–1998 (2018)
118. Tunali, E.T., Tarn, T.J.: New results for identifiability of nonlinear systems. IEEE Trans. Autom. Control **32**, 146–154 (1987)
119. Tuncer, N., Gulbudak, H., Cannataro, V.L., Martcheva, M.: Structural and practical identifiability issues of immuno-epidemiological vector-host models with application to Rift Valley Fever. Bull. Math. Biol. **78**, 1796–1827 (2016)
120. Tuncer, N., Le, T.T.: Structural and practical identifiability analysis of outbreak models. Math. Biosci. **299**, 1–18 (2018)
121. Tuncer, N., Marctheva, M., LaBarre, B., Payoute, S.: Structural and practical identifiability analysis of Zika epidemiological models. Bull. Math. Biol. **80**, 2209–2241 (2018)
122. Villaverde, A.: Observability and structural identifiability of nonlinear biological systems. Complexity **2019**, 1–13 (2019)
123. Villaverde, A., Banga, J.: Reverse engineering and identification insystems biology: strategies, perspectivesand challenges. J. R. Soc. Interface **11**, 1–16 (2014)
124. Walter, E., Lecourtier, Y.: Global approaches to identifiability testing for linear and nonlinear state space models. Math. Comput. Simul. **24**, 472–482 (1982)
125. Walter, E., Pronzato, L.: Identifiabilities and nonlinearities. In: Nonlinear Systems, vol. 1, pp. 111–143. Chapman and Hall, London (1995)
126. Walter, E., Pronzato, L.: Identification of Parametric Models. Communications and Control Engineering Series. Springer, Berlin (1997). From experimental data, Translated from the 1994 French original and revised by the authors, with the help of John Norton
127. Wang, Y., Sontag, E.D.: On two definitions of observation spaces. Syst. Control Lett. **13**, 279–289 (1989)
128. Wieland, F.G., Hauber, A.L., Rosenblatt, M., Tönsing, C., Timmer, J.: On structural and practical identifiability. Curr. Opin. Syst. Biol. **25**, 60–69 (2021)
129. Wonham, W.: Linear Multivariable Control: A Geometric Approach, 2nd edn. Springer, Berlin (1979)
130. Wu, H., Zhu, H., Miao, H., Perelson, A.S.: Parameter identifiability and estimation of HIV/AIDS dynamic models. Bull. Math. Biol. **70**, 785–799 (2008)
131. Xia, X.: Estimation of HIV/AIDS parameters. Autom. J. IFAC **39**, 1983–1988 (2003)
132. Xia, X., Moog, C.H.: Identifiability of nonlinear systems with application to HIV/AIDS models. IEEE Trans. Autom. Control **48**, 330–336 (2003)

Printed in the United States
by Baker & Taylor Publisher Services